FLOWS IN
TRANSPORTATION NETWORKS

This is Volume 90 in
MATHEMATICS IN SCIENCE AND ENGINEERING
A series of monographs and textbooks
Edited by RICHARD BELLMAN, *University of Southern California*

The complete listing of books in this series is available from the Publisher upon request.

Flows in Transportation Networks

RENFREY B. POTTS

Department of Applied Mathematics
University of Adelaide
Adelaide, Australia

ROBERT M. OLIVER

Department of Industrial Engineering
and Operations Research
University of California
Berkeley, California

 1972

ACADEMIC PRESS   *New York and London*

ACADEMIC PRESS, INC.
111 Fifth Avenue, New York, New York 10003

United Kingdom Edition published by
ACADEMIC PRESS, INC. (LONDON) LTD.
24/28 Oval Road, London NW1

LIBRARY OF CONGRESS CATALOG CARD NUMBER: 70-182673

PRINTED IN THE UNITED STATES OF AMERICA

CONTENTS

Preface   ix
Acknowledgments   xi

Chapter I TRANSPORTATION NETWORKS

1. Introduction   1
2. Examples of Transportation Networks   2
   (a) City Street Network   2
   (b) Main Road Network   4
   (c) Traffic Desire Network   9
   (d) Spider Web Network   9
3. Transportation Planning Process   10
4. Conclusion   13
5. Notes and References   13

Chapter II ELEMENTS OF NETWORK THEORY

6. Introduction   17
7. Graphs: Definitions and Notations   18
   (a) Directed Graph   18
   (b) Chain and Cycle   19
   (c) Path and Mesh   21
   (d) Accessible and Connected Nodes   22
   (e) Cut-Set   22
   (f) Undirected and Mixed Graphs   24
   (g) Tree and Arborescence   24

  8. Flows and Conservation Laws   26
      (a) Link Flows and Kirchhoff's Law   26
      (b) Single O–D Network: Link Flows   29
      (c) Single O–D Network: Chain Flows   32
      (d) Multiple O–D Network   34
      (e) Compressibility and Separability   36
  9. Costs and Capacities   38
      (a) Link, Route, and Network Costs   38
      (b) Capacitated Network   41
10. Conclusion   43
11. Notes and References   44
12. Problems   45

Chapter III EXTREMAL PRINCIPLES AND TRAFFIC ASSIGNMENT

13. Introduction   49
14. Cheapest Routes   51
      (a) Appraisal of Algorithms   51
      (b) Tree-Building Algorithms   52
      (c) Turn Penalties and Prohibitions   56
      (d) Cheapest Route Assignment   63
15. Minimum Network Cost   65
      (a) Link Flows   66
      (b) Chain Flows   71
      (c) The Out-of-Kilter Algorithm   75
16. Flow Dependent Costs   86
      (a) Multicommodity Formulation   87
      (b) Equilibrium Flow Patterns for Noncooperative Users   88
      (c) Minimum Network Cost Flow Patterns   91
      (d) Associated Traffic Assignment Problems   95
      (e) A Numerical Example with Four Commodities   96
      (f) Congested Assignment   100
17. Notes and References   102
18. Problems   109

Chapter IV TRIP DISTRIBUTION

19. Introduction   115
20. Model Formulation   116
21. Hitchcock Model   118
22. Entropy Models   121
      (a) Network Entropy   121
      (b) Proportional Model   123
      (c) Mean Trip Length Model   130
      (d) Gravity Model   133

23. Opportunity Models   136
    (a) Intervening Opportunities Model   137
    (b) Preferencing Model   138
24. Combined Distribution and Assignment   141
    (a) TRC Program   142
    (b) LTS Program   142
    (c) Multicommodity Distribution–Assignment   143
25. Conclusion   143
26. Notes and References   144
27. Problems   149

Appendix A   THEOREM FOR CHEAPEST ROUTE
             ALGORITHMS   153

Appendix B   DUALITY THEORY   157

Appendix C   INEQUALITIES FOR MARGINAL AND
             AVERAGE LINK AND CHAIN COSTS   159

Appendix D   ANSWERS TO PROBLEMS   163

Index   187

PREFACE

Transportation problems are among the most significant being faced by society today, and much effort is being and will be expended on the search for transportation systems which are efficient, acceptable to man, and compatible with his environment. Foremost in this search is the development of sophisticated mathematical models which are being used to analyze transportation problems and to plan for transportation needs of the future.

This text is designed to provide a comprehensive formulation of the more important transportation models; it purposely steers a middle course between theory *per se* on the one hand and applications without theorems on the other. Its aim is to bridge the gap between abstract graph theory and its application to the analysis of large transportation networks. The approach recognizes and emphasizes the ever increasing role of computer algorithms and, in doing so, selects those models and algorithms which, in the short period that they have been popularized in the open literature, appear to warrant long-term interest. The major portion of the text is based on fundamental conservation and extremal principles, and only seven statements have been classified as theorems.

Readers primarily interested in transportation planning will find that the book provides mathematical material necessary for proper understanding of the traffic assignment and trip distribution models widely used in the planning process. Science and engineering students will find that network flow theory is here motivated by applications to significant real transportation problems.

The book begins with an introductory chapter describing a variety of transportation networks, followed by a chapter summarizing flow concepts used in transportation problems. The third chapter is a detailed study of extremal principles, equilibrium flow patterns, and their application to tree-building algorithms and traffic assignment procedures. The final chapter is devoted to an analysis of the more important trip distribution models.

References at the end of each chapter are selective in quality without any attempt at being exhaustive in quantity, and together with the brief annotations, they should provide a reliable reference source and guide to the extensive literature on transportation models. The set problems, with their solutions collected in an appendix, should not only provide illustrations of the material but also a test of the reader's understanding of it.

Earlier versions of the manuscript have been used as a text for a graduate course for Master's degree students in Transportation at the University of California, and for graduate courses for students in Applied Mathematics and Operations Research at the University of California and the University of Adelaide. A shorter version was also successfully used in the University of California Extension program for a course given jointly by the authors to a mixed audience of engineers, architects, and planners from government and industry. The many constructive criticisms from a wide spectrum of colleagues and students, have helped to mould this text into its present form, and for this help the authors are especially grateful.

The following numbering conventions are used throughout the book. Sections have been numbered consecutively, 1 through 27, and the Appendixes, A through D. Equation and figure numbering begins anew with each section. In referring to an equation or figure, the section number or appendix letter is explicitly indicated only when it is in a section different from the current one. Numbers in square brackets correspond to references at the end of each chapter. Again, the appropriate section number is explicitly indicated only when the reference is at the end of a chapter different from the current one.

ACKNOWLEDGMENTS

The authors wish to thank Dr. Alan Goldman for his encouragement, support, and helpful criticisms, and Linda Betters and Norene Revelli for their excellent editorial and typing assistance. It is a pleasure to thank wives, families, students, graduates, colleagues, and the publisher, who have patiently watched, waited, and worked to help the authors complete an enjoyable task which required their closest cooperation— at a geographical separation of 8000 miles.

I

TRANSPORTATION NETWORKS

1. Introduction

The use of network models has been increasingly widespread in the recent rapid development of operations research. Because of the basic simplicity and generality of network concepts and because of the amenability of network calculations to digital computation, networks have proved ideally suited for mathematical modeling in a variety of scientific and engineering applications. Critical path methods, personnel assignment, job-lot scheduling, and flows in networks have become well established as fundamental operations research disciplines.

It is the purpose of this text to give an analysis of some of the basic problems and important applications involving flows in transportation networks. The use of network models has become universal in transportation planning for vehicular traffic. Although large transportation networks for many cities have been analyzed with considerable ingenuity, their usefulness to the planner has not been as great as it might have been, partly because the basic network concepts and optimization criteria have tended to become obscured and submerged in a mass of data and computer output. There is a need, which this text is designed to meet, for a clear statement of the basic principles underlying the theory and application of network flows in transportation problems. Although the emphasis will be on applications to vehicular traffic, the basic ideas described in this text can easily be applied to railroad, shipping, and airline networks.

In this chapter, the general scope of network problems will be illustrated by several examples of transportation networks, and these will be followed by a brief resume of the transportation planning process and the role played by network analysis. It is convenient to adopt the conventional description of a *transportation network* as a set of *nodes* with interconnecting *links*, and nodes will be represented by (numbered) circles and links by (arrowed) lines. This description anticipates the formal definitions, notations and terminology which will be introduced in detail in Chap. II.

2.   Examples of Transportation Networks

(a) City Street Network

The streets of a city afford an obvious example of a network. For a simple geographical representation of the street system, a *city street network* can be defined in which the nodes represent intersections and perhaps other important locations on roads, and the links represent the street segments. For the network to be useful in describing the movement of city traffic, it would be important to distinguish between one-way and two-way streets. A one-way street could be represented by a *directed link* with an arrow indicating the permitted direction of traffic movement. A two-way street could be represented by two directed links with arrows in opposite directions or by an *undirected link* without an arrow.

Figure 2.1 illustrates the city street network representing part of the street system of San Francisco, California, U.S.A. The network has 52 nodes (representing 32 street intersections, 17 points on the boundary of the area, and the extremities of 3 dead-end streets), 42 directed links (one-way streets) and 25 undirected links (two-way streets). Figure 2.2 is a simplified representation of the traffic movements at an intersection of two of the one-way streets (Kearny and Clay) in Fig. 2.1, in which node 7 representing this intersection is replaced by four nodes 71–74. Without traffic control, nodes 73 and 74 represent merge points and the crossing of links (71, 74) and (72, 73), a conflict point. For a two-phase traffic light, the conflict point is eliminated, with the A and B phase traffic movements as shown in Fig. 2.2(b).

A traffic engineer concerned with the control of traffic throughout a city would need a detailed description of all possible traffic movements

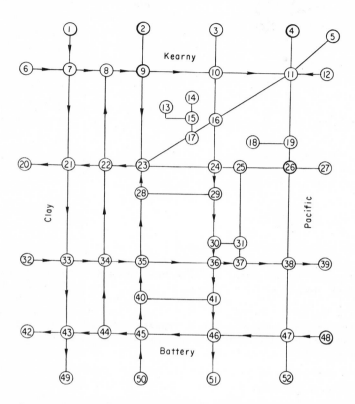

Figure 2.1. City street network representing part of the San Francisco street system bordered by Kearny, Pacific, Battery, and Clay Streets. The numbered nodes mostly represent intersections, the directed links one-way streets, and the undirected links two-way streets.

throughout the street system, and the corresponding network could be required to represent separate traffic lanes, turn prohibitions, turn lanes, allowed movements for various phases of traffic lights, and so on. The complexity of a city street network representation is indicative of the difficulty in attempting a detailed analysis or simulation of city traffic. Rather than considering the *microscopic* characteristics of traffic, this text will be primarily concerned with *macroscopic* traffic behavior and its interrelation with networks.

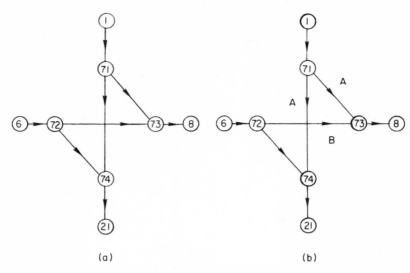

(a)   (b)

Figure 2.2.   Network giving a simplified representation of the traffic movements at the Kearny and Clay intersection. Node 7 of Fig. 2.1 is replaced by the four nodes 71–74. (a) Without traffic control, merging takes place at nodes 73 and 74 and links (71, 74) and (72, 73) cross at a conflict point. (b) The conflict point is eliminated by traffic light control with two phases represented by A, B. (Note: it has been assumed that traffic keeps to the right and that a right, but not a left, turn against a red signal after a stop is allowed.)

(b) Main Road Network

To study the macroscopic movement of traffic throughout an extensive region, it is usual to divide the area into subareas and to concentrate attention on the main roads. The subdivision of the study area may proceed in stages. First the area is divided into a few *sectors*, then into smaller *districts*, and finally into *zones*. The zones are the basic subareas used in transportation studies and are chosen so that each has reasonably uniform land use characteristics. The traffic origins and destinations throughout a zone are assumed concentrated at a point which is represented by a special node, called a *centroid*. Each centroid is connected to the main road system by one or more *dummy links*. The main road system, together with the centroids and dummy links, forms a *main road network*. The nodes of a main road network are of two kinds— *intermediate nodes* representing main intersections, and centroids. The links are also of two kinds—links representing the main roads and dummy links. A convenient convention is to distinguish centroids as double circles and dummy links as dashed lines.

It is important to realize that in constructing or, as a transportation planner would say, *coding* a main road network, only a gross representation of the actual road system is required. Several main parallel roads may be represented by a single link and a complex of intersections by a single intermediate node. To achieve simplicity, most, if not all, links may be shown to be undirected, all turns (except U-turns) allowed at intermediate nodes, and conflict points eliminated so that links do not cross. These restrictions scarcely reduce the generality of the representation. Even if a turn at a major intersection were actually prohibited, traffic wishing to interchange could usually do so by using nearby minor roads. To study the gross traffic movements throughout an extensive region, a main road network is adequate, but it has to be coded and interpreted with care.

In transportation studies it is often convenient to use main road networks of varying complexity. For the Bay Area Transportation Study (BATS), the study area encompasses nine counties surrounding the San Francisco and San Pablo Bays (see Fig. 2.3 and also [1]). For purposes of the study, BATS has used two main road networks, a sketch network with about 1000 nodes (including 300 centroids) and 1400 undirected links, and a detailed network of about 4000 nodes (including 1200 centroids) and 5500 links (mostly undirected). As an example of a main road network for use in later chapters, we illustrate in Fig. 2.4 an extremely gross main road network with 9 centroids representing the counties, 15 intermediate nodes, and 30 undirected links (including 12 dummy links).

In using a main road network, a transportation planner associates various parameters with the links and nodes. For example, to each link may be attached values indicating the number of traffic lanes, the road length, the average travel time, the average vehicle speeds, the average daily traffic flow, the peak hour flows, and the capacity. To each intermediate node may be associated time penalties for left and right turn movements, and to each centroid the flow of traffic assumed to originate and terminate there. Separate parameters may be used for various traffic stratifications corresponding to different modes (truck, private automobile, transit) and to different trip purposes (home-to-work, shopping, social, recreation, school).

In measuring traffic flows on a main road network, it is often convenient for checking purposes to count traffic crossing screen and

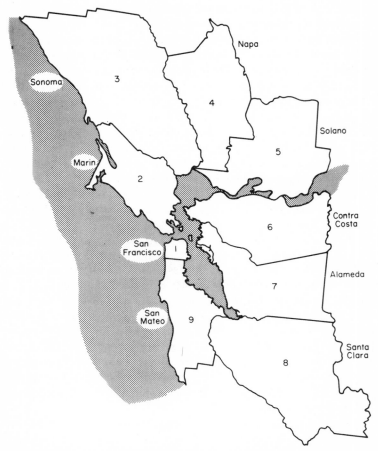

Figure 2.3.   Study area for the Bay Area Transportation Study (BATS), California, U.S.A. The nine counties forming the study area are (1) San Francisco, (2) Marin, (3) Sonoma, (4) Napa, (5) Solano, (6) Contra Costa, (7) Alameda, (8) Santa Clara, and (9) San Mateo.

cordon lines. A *screen line* completely separates two subareas of the study area, and the traffic is counted on the main roads where it crosses the screen line. A *cordon line* is similar to a screen line except that it completely encloses a subarea. In choosing screen and cordon lines, use is made of natural boundaries, such as rivers, bays, or mountains, so that the screen line and cordon counts can be measured easily and accurately. From Fig. 2.3 it is evident that some of the county borders would be suitable as cordon and screen lines for the Bay Area. It is interesting to note that the general concept of screen lines plays an important role in the general theory of flows in networks.

It is pertinent to point out that if no consideration were being given

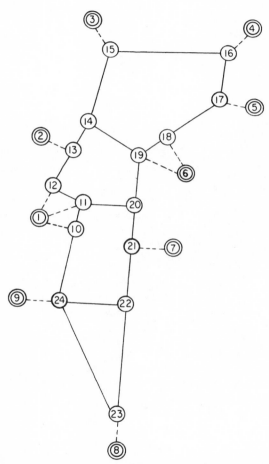

Figure 2.4. An example of a gross main road network for the Bay Area. The nine centroids (double circles) represent the counties and are connected to the main roads by dummy links (dashed lines). Five bridges are represented by the following links: (11, 20) Bay Bridge; (12, 13) Golden Gate; (14, 19) Richmond–San Rafael; (18, 19) Carquinez Straits; (22, 24) San Mateo.

to the traffic flows, capacities, and other parameters, the collection of
nodes and links representing the topology of the main road system
would afford an example of what is usually referred to in the literature
as a *graph* (see Sect. 7). The more specific terms, network and transportation network, are used when the movement of traffic throughout
the system is being analyzed.

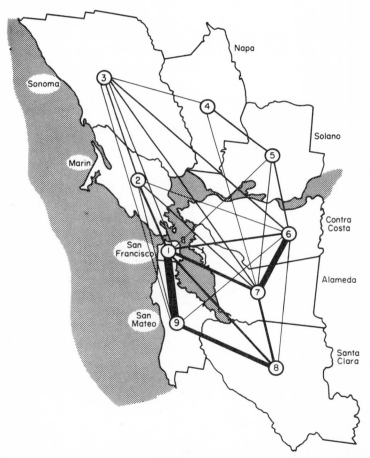

Figure 2.5. A traffic desire network for the Bay Area. The nodes represent the
counties (see Fig. 2.3) and the links the intercounty traffic desires. The widths of
the desire lines are proportional to the total intercounty private automobile trips on
an average weekday in 1965.

(c) Traffic Desire Network

In analyzing the traffic problems in an extensive region, it is customary to describe the traffic movements as trips between origins and destinations. Trips between all points in a traffic zone and all points in another zone may be aggregated to give the traffic desire between the two zones, which can be represented by a directed or undirected desire line, a straight line joining the centroids of the zones. In the traffic desire network, the nodes represent the centroids and the links the desire lines. Although the nodes of the traffic desire network have geographical significance, the links do not represent roads or traffic routes and the crossings of desire lines have no significance (see Fig. 2.5). Because the network has no intermediate nodes, it is unnecessary to use double circles to represent the centroids. Associated with each link is the traffic desire, measured, for example, by the average number of weekday interzonal trips and often represented pictorially by a desire line of width proportional to the traffic desire.

The traffic desire network is an example of a *complete nonplanar* network: complete because each pair of nodes is joined by a link (or possibly two directed links), and nonplanar because it cannot be represented on a plane with noncrossing links (except when the number of nodes is less than five). It is a useful representation of the traffic desires only if the number of zones is small, as it rapidly becomes complicated as the number of nodes increases. Then it is more convenient to represent the traffic desires graphically by a spider web network or algebraically by a trip table or distribution matrix (see Chap. IV).

(d) Spider Web Network

For a region with a large number of zones, it is not convenient to accumulate interzonal trips onto desire lines between all pairs of zones, but it is better to assign the trips to a spider web network consisting of nodes representing the centroids and links representing desire lines between adjacent zones. Because its links do not cross, the spider web network is a planar network. In assigning an individual zone-to-zone desire to the spider web network, the choice of a route from the origin node to the destination node *via* adjacent nodes has to be based on some criterion such as shortest distance. The spider web network with the accumulated traffic desires is useful as it exhibits the general features of the traffic flow patterns.

Figure 2.6 is a spider web network obtained by assigning the inter-county desires illustrated in Fig. 2.5 to a very crude network representing the main intercounty connections. The reader is referred to [2] for some excellent illustrations of large spider web networks.

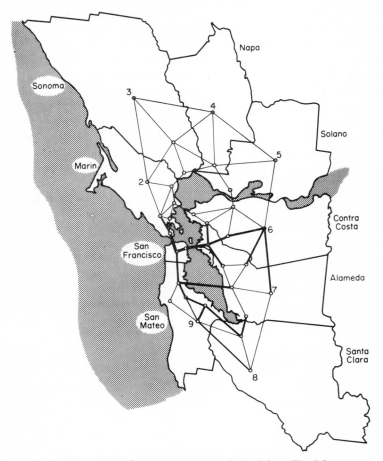

Figure 2.6.   Spider web network obtained from Fig. 2.5.

3.   Transportation Planning Process

The networks which have been described above are commonly used in transportation studies. As an analysis of their properties is one of the main purposes of this text, it is important to summarize briefly the various phases of the transportation planning process.

The broad objective of a transportation study of a metropolitan area is the development of a comprehensive long-range transportation plan. This requires the estimation of future automobile and transit traffic and the assessment of an efficient and economical transportation system to serve the predicted travel patterns. It is customary to base this extremely formidable task on the concept of mathematical models. The general philosophy is that there is a regularity in the habits of an urban population which establishes certain patterns in movements of the people and their goods. These patterns are detected by the systematic collection and inventory of transportation data, and they are described by mathematical models involving various constants and other parameters related to social and economic characteristics of the population, and the location of various activities throughout the study area. The mathematical models are tested against traffic patterns measured for some "base" year and are "calibrated" by choosing the model constants to give a best fit. The models are then used to forecast the future travel conditions for one or more "design" years with assumed and predicted new values of the model parameters. Alternative transportation plans are tested and evaluated to determine which will best meet the anticipated demands.

The validity of the transportation planning process has yet to be established. The approximate reproduction of the base year traffic patterns has proved possible, although the models tend to lose their basic mathematical structure when it proves necessary to introduce "fudge" factors to give a satisfactory fit to the data.

The vital phase of using the models to predict future traffic patterns—peak, average daily travel, automobile, transit—has yet to be adequately tested, and some preliminary checking of cities with transportation plans which have now reached their design years has forced the conclusion that traffic prediction is still more an art than a science!

To describe the important role of networks in transportation planning, it is convenient to distinguish four phases of the planning process, and to highlight (with an asterisk) those aspects which are most amenable to network analysis and which will be described and discussed in detail in the subsequent chapters as indicated.

Phase I—Base Year Inventory

The first phase of the process can be conveniently separated into three separate inventory tasks:

*(i)   *Inventory of main roads and transit services* (Chaps. II, III)
Definition of study area and division into sectors, districts, and zones.
Coding of city street, main road, and transit networks.
Measurement of traffic flows, speeds, travel times, and link lengths and capacities.

*(ii)   *Inventory of travel patterns* (Chap. II)
Determination of intrazonal and interzonal traffic desires by origin–destination (O–D) studies, screen line and cordon counts.
Drawing traffic desire and spider web networks.

(iii)   *Inventory of planning factors*
Land use, housing, employment, industry, business, shops, schools, recreation.
Vehicle registrations, car ownership, household income.

Phase II—Model Analysis

The second phase of the process is the determination and base-year calibration of the following mathematical models:

(i)   *Trip generation*
Mathematical model relating planning factors to the origins and destinations of person or vehicle journeys (or to the so-called production of trips from zones and the attraction of trips to zones).

*(ii)   *Trip distribution* (Chap. IV)
Mathematical model relating intrazonal and interzonal trips to trip generation.

*(iii)   *Traffic assignment* (Chap. III)
Mathematical model allocating interzonal trips to the main road and transit networks.

Phase III—Travel Forecasts

In the third phase of the process, the design year travel is forecasted using the predicted planning factors and the calibrated mathematical models with the estimated model parameters. The transportation system forms a feed-back from the next phase into this phase; the two phases are repeated until a satisfactory future system is selected.

Phase IV—Network Evaluation

In the final phase of the process, alternative future transport systems are evaluated and the preferred one selected:

    (i)   *Proposed transport systems*
         Various future systems are proposed, formulated, and represented in the forms required by the models.

  *(ii)   *Future traffic assignment* (Chap. III)
         Assignment of forecast travel to proposed road and transit networks.

  (iii)   *Future traffic analysis*
         Traffic flow and capacity analysis of forecast travel patterns in relation to design standards.

  (iv)   *Economic analysis*

A more detailed description of these phases is outside the scope of this text, but the interested reader will find references [1]–[4] complete and comprehensive.

4. Conclusion

The examples mentioned in this chapter give some indication of the variety of networks occurring in transportation problems. Their use in transportation planning has initiated considerable research into computer techniques for coping with intricate networks and, as is often the case, the mathematicians have already provided in graph theory a well-developed framework on which to build the theory of flow in transportation networks. It is to the basic theory that we turn in the next chapter.

5. Notes and References

The published reports on the numerous transportation studies which have been carried out for major cities throughout the world provide an excellent source of material on transportation networks. Most of the reports are lavishly produced with an abundance of tables, diagrams, and illustrations of networks, but rarely do they detail technical and mathematical aspects of the study methods adopted.

The following three references are representative of the material available:

[1]   Bay Area Transportation Study Commission, *Study Design*, 1966.

This concise report outlines the background, objectives, and general concepts of BATS and summarizes the plan of the study and its timing and financing. The design of the study seems to have proved rather ambitious, and significant modifications have been proved necessary for the practical implementation of the planned program.

As in a number of studies, BATS has produced many important specialized reports on various aspects of transportation planning.

[2]   Greater London Council, *London Traffic Survey*, Vol. 1 (1964), Vol. 2 (1966), and *Movement in London* (1969).

These three volumes represent some of the best technical knowledge of transportation planning in the U.S.A. and England applied to one of the world's greatest cities. Volume 1 contains an almost interminable mass of data on the travel habits of nearly nine million persons and effectively covers Phases I and II of the transportation planning process as described above. Volume 2 is concerned with detailed forecasting of the extent and nature of future travel demands. *Movement in London* is in effect the third volume describing the survey with emphasis on research and policy aspects. The three volumes contain excellent illustrations of main road networks, traffic desire networks, spider web networks, and other transportation networks.

[3]   Metropolitan Corporation of Greater Winnipeg, *Winnipeg Area Transportation Study*, Vols. 1, 2 (1966), Vol. 3 (1968).

This study is an interesting example of the application of the planning process to a smaller city. The reports contain considerable technical detail of the mathematical basis of the study and again provide some excellent illustrations of transportation networks.

A more accessible reference is the following:

[4]   Smith, Wilbur S., *Manual on Urban Planning. Chapter VII: Transportation Planning*, Journal of the Urban Planning and Development Division, Proceedings of the American Society of Civil Engineering, Vol. 93, No. UP2, pp. 93–143 (1967).

This comprehensive paper, by one of the leading American transportation planners, gives a detailed account of the planning process with full explanations of the terminologies and jargon peculiar to planners. The author presents an interesting comparison of the studies which he has carried out in various American cities.

II

ELEMENTS OF NETWORK THEORY

6. Introduction

It is the purpose of this chapter to give a brief outline of the elements of the theory of graphs and transportation networks, emphasizing those aspects which are most relevant to the applications considered in later chapters. This requires a rather unusual emphasis on some sections of the standard theory, but enables much to be excluded. From the great variety of confusing terminology and notation in the literature [1]–[6], we have tried to select that notation which is most appropriate for the transportation problems we discuss. Some of the terminology has already been used in the previous chapter and in Sect. 7 this terminology and corresponding notation is rigorously defined. The general term *route* includes both the special case when an enroute traveler has to follow link directions and the case when this is not necessary. Section 8 introduces the concept of link flows and the conservation equations they must satisfy; these are first illustrated for networks with a single origin–destination pair and then extended to multiple O–D pairs. The important concepts of link, route, and network costs are introduced in Sect. 9, together with a description of the properties of capacitated networks.

7. Graphs: Definitions and Notations

(a) Directed Graph

A *directed graph* $[N; L]$ is defined as a finite set N of unordered elements and a set L of ordered pairs of elements of N. We denote by n and l the numbers of elements in the sets N and L, respectively.

The elements of N are called *nodes* and are denoted by i or n_i, $i = 1, 2, ..., n$. The elements of L are called *links*, or more specifically *directed links*, and are denoted by (i, j) or (n_i, n_j). Alternatively, the links can be enumerated as i or $l_i, i = 1, 2, ... , l$. The latter notation is necessary when we wish to allow for two or more parallel links between the same two nodes; in general, we exclude parallel links unless specifically stated to the contrary.

The two nodes defining a link may or may not be distinct. If the nodes are the same, then the link is called a *loop* and is denoted by (i, i) or (n_i, n_i).

A link (i, j) is said to *join* the nodes i and j, and it is common practice in transportation applications to call i the A-node and j the B-node[†] of the link (i, j).

We shall occasionally use the terms partial graph and subgraph. A *partial graph* of a directed graph $[N; L]$ is a directed graph $[N; L']$ with L' a subset of L, i.e., $L' \subseteq L$. A *subgraph* of $[N; L]$ is a directed graph $[N'; L']$ with $N' \subseteq N$ and where L' is the set of all links of L which join nodes of N', i.e.,

$$L' = \{(i, j)|(i, j) \in L, i \in N', j \in N'\}. \tag{1}$$

A partial graph can be obtained from a directed graph by deleting links, and a subgraph by deleting nodes and attached links.

Two special graphs are of particular importance in network applications. The first is the so-called *complete graph*, for which there is at least one link joining any two distinct nodes of N, i.e.,

$$i \in N, \quad j \in N, \quad i \neq j, \quad (i, j) \notin L \Rightarrow (j, i) \in L. \tag{2}$$

The second special graph is a *bipartite graph* in which the set N of nodes is partitioned into two complementary sets $X, \overline{X}$, i.e.,

$$X \cup \overline{X} = N, \quad X \cap \overline{X} = \varnothing = \text{empty set}. \tag{3}$$

---

†This notation should not be confused with that for "after" and "before" nodes introduced in Sect. 8.

The set L is the set of links joining nodes of X to nodes of $\bar{X}$, i.e.,

$$L = \{(i,j)|i \in X, j \in \bar{X}\}. \tag{4}$$

It should be noted that a bipartite graph cannot contain any loops.

The following simple example illustrates these concepts. For the directed graph $[N;L]$ with

$$N = \{1,2,3,4\}, \tag{5}$$

$$L = \{(1,2),(1,3),(1,4),(2,3),(2,4),(3,2),(3,3),(3,4)\}, \tag{6}$$

(i)   1 is the A-node and 2 the B-node of the directed link $(1,2)$;

(ii) $[N;L']$ is a partial graph with

$$L' = \{(1,2),(1,3)\}; \tag{7}$$

(iii) $[N';L']$ is a subgraph with

$$N' = \{(1,2,3)\}, \tag{8}$$

$$L' = \{(1,2),(1,3),(2,3),(3,2),(3,3)\}. \tag{9}$$

The graph $[N;L]$ is a complete graph but not bipartite. A simple example of the latter is given by

$$N = \{1,2,3,4\}, \qquad X = \{1,2\}, \qquad \bar{X} = \{3,4\}, \tag{10}$$

$$L = \{(1,3),(1,4),(2,3),(2,4)\}. \tag{11}$$

These algebraic definitions are more readily understood when the geometrical representations of graphs are drawn. A node is represented by a numbered circle and a directed link by an arrowed line. The two diagrams in Fig. 7.1 illustrate that a particular geometrical representation of a graph need have no geographical significance. However, for applications to transportation problems, geographical considerations are of great importance.

(b) Chain and Cycle

In determining routes through networks, we shall find it important to distinguish between those for which the link directions have to be followed and those for which this is not necessary. The contrast is between the motorist who has to obey one-way street signs and the pedestrian who can ignore them. The pedestrian has a wider choice of routes because he can walk the wrong way down a one-way street or along the wrong side of a two-way street.

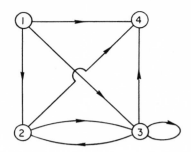

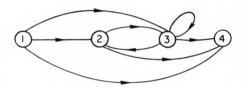

Figure 7.1.   Two geometrical representations of the same directed graph.

When the directions are important, the routes are called chains and cycles. A chain of a directed graph is formally defined as follows: If $n_1, n_2, ..., n_r$ are distinct nodes and $(n_i, n_{i+1}), i = 1, 2, ..., r-1$ are directed links, then the sequence

$$n_1, (n_1, n_2), n_2, ..., n_{r-1}, (n_{r-1}, n_r), n_r \tag{12}$$

defines a *chain* from the *origin node* n_1 to the *destination node* n_r. For the directed graph illustrated in Fig. 7.1,

$$1, (1,4), 4 \tag{13}$$

and

$$1, (1,2), 2, (2,3), 3, (3,4), 4 \tag{14}$$

are chains from origin node 1 to destination node 4. They represent allowable routes for a motorist.

It is clear that the sequence of nodes or the sequence of links is sufficient to define a chain uniquely. Thus, the chain in (14) could be adequately denoted by

$$1, 2, 3, 4 \tag{15}$$

or

$$(1,2), \quad (2,3), \quad (3,4).\tag{16}$$

It will be sometimes convenient to denote chains by j or $m_j, j = 1, 2, \ldots, m$, and to denote a set of chains by M.

A *cycle* is defined as a chain, except that $n_1 = n_r$, and is an allowable there-and-back route or round trip for a motorist. For the graph in Fig. 7.1,

$$2, \quad (2,3), \quad 3, \quad (3,2), \quad 2\tag{17}$$

is a cycle; it could equally well be denoted by

$$3, \quad (3,2), \quad 2, \quad (2,3), \quad 3\tag{18}$$

as it is immaterial which is the starting and which is the finishing node.

It is important to note that the nodes of a chain are required to be distinct. This means that a chain cannot contain a cycle or a loop.

(c) Path and Mesh

When the directions of links do not have to be followed, the routes through a graph are called paths and meshes, analogous to the chains and cycles already defined. The formal definition of a path is as follows: if $n_1, \ldots, n_i, \ldots, n_r$ are distinct nodes, and (n_i', n_{i+1}'') are links, then a *path* from the origin node n_1 to the destination node n_r is defined by the sequence

$$n_1, \quad (n_1', n_2''), \quad n_2, \ldots, n_i, \quad (n_i', n_{i+1}''), \quad n_{i+1}, \ldots, n_r,\tag{19}$$

where either $n_i' = n_i$ and $n_{i+1}'' = n_{i+1}$, in which case (n_i', n_{i+1}'') is called a *forward link* of the path, or else $n_i' = n_{i+1}$ and $n_{i+1}'' = n_i$, in which case (n_i', n_{i+1}'') is a *reverse link* of the path. For the graph illustrated in Fig. 7.1,

$$1, \quad (1,4), \quad 4, \quad (3,4), \quad 3, \quad (2,3), \quad 2\tag{20}$$

is a path from origin node 1 to destination node 2 with $(1,4)$ a forward link and $(3,4)$ and $(2,3)$ reverse links. Another path through the same nodes in the same order is

$$1, \quad (1,4), \quad 4, \quad (3,4), \quad 3, \quad (3,2), \quad 2,\tag{21}$$

with $(3,4)$ the only reverse link. It is clear that a sequence of nodes does not uniquely define a path, whereas a sequence of directed links does. Thus,

$$1, 4, 3, 2\tag{22}$$

does not distinguish between the two paths (20) and (21), but

$$(1,4), \quad (3,4), \quad (2,3) \tag{23}$$

uniquely defines the path (20).

A *mesh* is defined as a path, except that $n_1 = n_r$. Thus,

$$2, \quad (2,3), \quad 3, \quad (3,4), \quad 4, \quad (2,4), \quad 2 \tag{24}$$

or

$$3, \quad (3,4), \quad 4, \quad (2,4), \quad 2, \quad (2,3), \quad 3 \tag{25}$$

are meshes of the graph in Fig. 7.1 with $(2,4)$ a reverse link. It is necessary to exclude from this definition the degenerate sequence

$$n_1, \quad (n_1, n_2), \quad n_2, \quad (n_1, n_2), \quad n_1, \tag{26}$$

which represents, in effect, simply a single link joining two nodes.

(d) Accessible and Connected Nodes

The existence of a chain from n_1 to n_r does not imply the existence of a chain from n_r to n_1; but the existence of a path from n_1 to n_r implies the existence of a path from n_r to n_1. A node n_r is said to be *accessible* from a distinct node n_1 if and only if there exists a chain from n_1 to n_r. Distinct nodes n_1 and n_r are said to be *connected* if and only if there exists a path from n_1 to n_r. Two distinct nodes may be connected but either one or both may be inaccessible from the other. A pedestrian can walk from a node to any other node connected to it, but a motorist can only drive to accessible nodes.

For a *connected directed graph*, all pairs of distinct nodes are connected. The two networks illustrated in Fig. 7.2 are both connected directed graphs. The first has "good accessibility" as each node is accessible to every other node. The second has poor accessibility and would be a motorist's nightmare!

(e) Cut-Set

If the set of nodes N is partitioned into complementary sets $X, \overline{X}$, then the subset of L defined by

$$(X, \overline{X}) = \{(i,j) | (i,j) \in L, i \in X, j \in \overline{X}\} \tag{27}$$

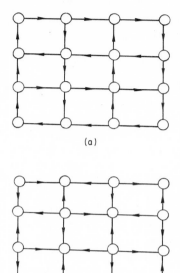

(a)

(b)

Figure 7.2. Two connected directed graphs. (a) Good accessibility, (b) poor accessibility.

is called a *cut-set*. We emphasize the fact that a cut-set is a subset of directed links. For the graph illustrated in Fig. 7.1,

$$(X, \overline{X}) = \{(1,3), (1,4), (2,3), (2,4)\} \tag{28}$$

and

$$(\overline{X}, X) = \{(3,2)\} \tag{29}$$

are cut-sets with $X = \{1,2\}$ and $\overline{X} = \{3,4\}$. Note that the notation $(X, \overline{X})$ is a generalization of that for a link (i,j).

The notion of a cut-set is basic to screen and cordon lines used in transportation studies. For example, if an East–West river is used as a screen line, X can be chosen as the set of nodes North of the river, and the cut-set $(X, \overline{X})$ is the set of all southbound roads on bridges crossing the river. A count of traffic on these roads would enable the total traffic from North to South of the river to be calculated.

If a cordon line is a ring road encircling the Central Business District of a city, X can be chosen as the set of nodes within the CBD, $\overline{X}$ the set of nodes outside, and $(X, \overline{X})$ is the set of roads crossing the ring road outwards from within the CBD. A traffic count on these roads at "external" stations on the ring road would enable the total traffic from within the CBD outwards to be calculated.

(f) Undirected and Mixed Graphs

For an *undirected graph* $[N; L]$, the elements of L are *unordered* pairs of elements of N and are denoted by (i,j) or (j,i). They are called *undirected links* and arrows are not needed in their geometrical representation. For an undirected graph, there is no distinction between the notions of chains and paths, cycles and meshes, accessible and connected nodes. The definition above for a cut-set remains the same for an undirected graph, but there is now no distinction between $(X, \overline{X})$ and $(\overline{X}, X)$.

In transportation studies, undirected graphs are used whenever possible in preference to directed graphs because of their simplicity. An undirected link of a transportation network allows two-way traffic and can always be regarded as two oppositely arrowed directed links. This is essential when cut-sets are used for cordon and screen lines as the directions of movement across the lines must be considered separately. It is important in applications to distinguish carefully between undirected and directed links. Too often confusion arises in traffic studies because flows on roads are quoted without indicating whether one or two way traffic is being considered.

Occasionally graphs are used in which some links are directed and others are undirected; such a graph is referred to as *mixed graph*. The city street network illustrated in Fig. 2.1 (Chap. I) is an example of a mixed graph.

(g) Tree and Arborescence

A *tree*, denoted by $[N; T]$, is a connected graph which has no meshes. This definition implies that link directions can be ignored when deciding whether a graph is a tree.

A graph is a tree if and only if every pair of distinct nodes is connected by precisely one path, since any mesh can be regarded as two alternative paths between two nodes.

A *spanning tree* of a graph $[N; L]$ is a tree $[N; T]$ which is a partial graph of $[N; L]$, i.e., $T \subseteq L$. For example, Fig. 7.3(b), (c), (d) are spanning trees of the graph illustrated in Fig. 7.2(a), whereas Fig. 7.3(a) is a tree which is not a spanning tree of this graph.

For any connected graph which is not a tree, it is possible to remove certain links in meshes without destroying the connectivity, although not every link in a mesh is a candidate for removal. This process is continued until a spanning tree is obtained. If a link were removed from the tree, a disconnected graph would be obtained, since any link is the unique path connecting the two nodes it joins. A spanning tree may, therefore, be characterized as a minimal connected graph, in the sense that it contains no proper connected partial graph.

Spanning trees are important in transportation applications when shortest paths in networks are determined. The shortest paths from any origin or *home node* to all other nodes of the network (assumed connected) together form a spanning tree. Any tie between paths of equal length is supposedly resolved by the arbitrary choice of one of the paths. The home node of such a shortest path tree is sometimes called the root of the tree.

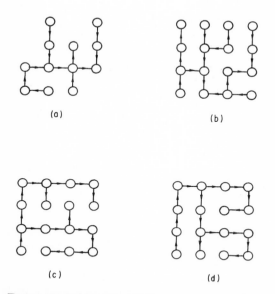

(a)

(b)

(c)

(d)

Figure 7.3. Trees and arborescences. (a) Tree which does not span the directed graph in Fig. 7.2(a); (b) spanning tree, but not an arborescence; (c), (d) spanning trees which are arborescences with the bottom left-hand node as home node.

When the directions of the links have to be taken into account, the tree which consists of chains from a home node to all other nodes is called an *arborescence*. For example, Fig. 7.3(c), (d) represent arborescences with the bottom left-hand node as a home node, while Fig. 7.3(b) represents a spanning tree which is not an arborescence for any node. Spanning trees of an undirected graph are arborescences with any node as home node.

8.   Flows and Conservation Laws

(a) Link Flows and Kirchhoff's Law

In many of the problems which we shall study, flows of vehicles, goods, or pedestrians can be associated with links of a graph; we then refer to the graph as a *network*, or more specifically as a *transportation network*, if the application to transportation is to be emphasized. The term *flow* denotes quantity per unit time, such as vehicles per hour, person trips per week day, or pedestrians per minute, and thus has the dimensions of a rate.

Fundamental to the theory of flow of electric currents in electrical networks, water in pipe networks, or traffic in transportation networks, is *Kirchhoff's law*, which is a conservation law stating that, for steady or static conditions, flows are neither created nor destroyed. The steady conditions imply for traffic applications that we are not concerned with the microscopic and stochastic characteristics of a traffic stream of individual vehicles travelling at random or in platoons on a city street network, but rather with the gross macroscopic behavior of traffic as, for example, on a main road network. We ignore fluctuations over time.

The interpretation of Kirchhoff's law for a node of a transportation network depends on whether or not the node produces or attracts traffic. For example, the intermediate nodes of a main road network merely serve as *enroute* points where travelers can merge and select alternative routes. In this case, Kirchhoff's law states that the sum of all flows leaving an intermediate node equals the sum of all flows entering the node. On the other hand, a centroid represents a zone where vehicle trips are *produced* by residents going on trips elsewhere and where trips are *attracted* to its places of employment. Kirchhoff's law then states that the sum of all flows leaving the centroid equals the flow produced at the centroid, and the sum of all flows entering the centroid equals the flow attracted to the centroid.

We shall adopt for a general transportation network the terminology of centroids and intermediate nodes to distinguish between nodes where traffic may be, and may not be, produced or attracted. In many other applications, the centroids are called *sources* and *sinks*. We shall adopt the following notation. The *link flow* on the directed link (i,j) will be denoted by f_{ij}, the flow *produced* at a centroid i by a_i, and the flow *attracted* to a centroid i by b_i. The quantities f_{ij}, a_i, b_i are assumed to be nonnegative. It is convenient to define $A(i)$ and $B(i)$, the set of nodes "after" and "before" node i by[†]

$$A(i) = \{j \,|\, j \in N, (i,j) \in L\}, \tag{1}$$

$$B(i) = \{j \,|\, j \in N, (j,i) \in L\}. \tag{2}$$

Kirchhoff's law for a directed transportation network $[N; L]$ can then be written in the form of *conservation equations* as follows:

$$\left.\begin{aligned}\sum_{A(i)} f_{ij} &= a_i, \\[6pt] \sum_{B(i)} f_{ji} &= b_i,\end{aligned}\right\} \quad \text{if } i \text{ is a centroid,} \qquad \begin{aligned}&(3)\\[12pt]&(4)\end{aligned}$$

$$\sum_{A(i)} f_{ij} - \sum_{B(i)} f_{ji} = 0, \qquad \text{if } i \text{ is an intermediate node.} \quad (5)$$

For these equations to have solutions, the total production, $\sum a_i = v$ say, must be equal to the total attraction $\sum b_i$. Since the number of links is generally at least twice the number of nodes in a network, the number of unknowns in Eqs. (3)–(5) greatly exceeds the number of equations and the equations are rich in solutions.

Figure 8.1 illustrates a transportation network with two centroids and two intermediate nodes. For the intermediate node 2, Kirchhoff's law can be easily verified:

$$i = 2, \qquad A(2) = \{3,4\}, \qquad B(2) = \{1,3\},$$

$$\sum_{A(2)} f_{2j} - \sum_{B(2)} f_{j2} = f_{23} + f_{24} - f_{12} - f_{32} = 5 + 1 - 3 - 3 = 0.$$

†This notation should not be confused with the use of A-node and B-node for the pair of nodes defining a directed link in Sect. 7, Chap. II.

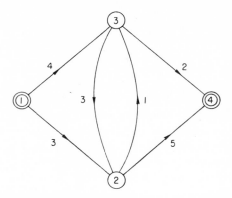

Figure 8.1.   Link flows on a network. Node 1 is a centroid, production=7, attraction=0. Node 4 is a centroid, production=0, attraction=7. Nodes 2, 3 are intermediate nodes, production—attraction=0.

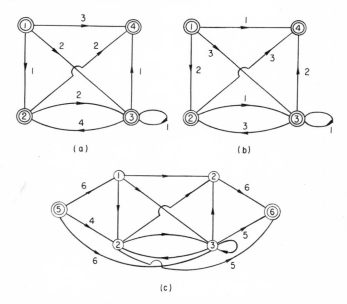

(a)   (b)

(c)

Figure 8.2.   Network with four centroids with given productions and attractions. (a), (b) Two different possible link flows; (c) transformed network with two centroids, four intermediate nodes, and flow value $v=16$.

node	i	1	2	3	4
production	a_i	6	4	6	0
attraction	b_i	0	5	5	6

For node 3 in Fig. 8.2(a), Kirchhoff's law can again be easily verified:

$$i = 3, \qquad a_3 = 6, \qquad b_3 = 5,$$

$$A(3) = \{2, 3, 4\}, \qquad B(3) = \{1, 2, 3\},$$

$$\sum_{A(3)} f_{3j} = f_{32} + f_{33} + f_{34} = 4 + 1 + 1 = 6 = a_3,$$

$$\sum_{B(3)} f_{j3} = f_{13} + f_{23} + f_{33} = 2 + 2 + 1 = 5 = b_3.$$

As a simple illustration of the nonuniqueness of the solutions of the conservation equations, the first two diagrams in Fig. 8.2 give two sets of link flows for the same productions and attractions at the centroids for a transportation network with four centroids and no intermediate nodes.

We shall first consider networks with just one centroid producing flow and one centroid attracting flow, and then extend our analysis to networks with more centroids. It is important to note that two or more nodes $i, j, \dots$ producing flow $a_i, a_j, \dots$ can often be replaced by an additional "fictitious" node with new links to nodes $i, j, \dots$ on which the flows are forced to be $a_i, a_j, \dots$. Similarly, two or more nodes attracting flow may be replaced by links to an additional "fictitious" attraction node. The flow value for this new network with the additional nodes and links is simply v, the total attraction or production. (See, for example, Fig. 8.2(c).) This replacement is always possible with single-commodity or single-copy flows but may not be possible for networks carrying multicommodity or multicopy flows of the type discussed in Sect. 8(d). For theoretical purposes, the transformation of a transportation network with many centroids to a network with just two centroids—one producing, the other attracting flow—can be of great importance, even though the additional nodes and links have no geographical significance.

(b) Single O–D Network: Link Flows

In a main road network there are many centroids acting as producers and attractors of traffic. In transportation studies, the traffic on such a network is analyzed as a superposition of traffic between specific origin-destination (O–D) pairs. If all other traffic except that between the O–D pair under question is ignored, the main road network becomes a *single* O–D *network* with two centroids, one the origin and the other the destination.

It will be convenient to let node $i = 1$ be the origin and node $i = n$ be the destination of a single O–D network, and to suppose the origin has zero attraction and production $g > 0$, and the destination zero production and attraction g. The quantity g is called the *flow value* of the network. Note that we specifically exclude the possibility that traffic leaving the origin might get lost and return to the origin! The conservation equations for such a network are

$$\sum_{A(1)} f_{1j} = g, \tag{6}$$

$$\sum_{A(i)} f_{ij} - \sum_{B(i)} f_{ji} = 0, \qquad i = 2, \ldots, n - 1, \tag{7}$$

$$- \sum_{B(n)} f_{jn} = -g. \tag{8}$$

Figure 8.1 is an example of a single O–D network with flow value 7.

For a single O–D network, we now prove the important

THEOREM: *The net flow across any cut-set separating the origin and destination is equal to the flow value.*

Proof: Any cut-set $(X, \overline{X})$ with origin $1 \in X$ and destination $n \in \overline{X}$ is said to separate the origin and destination. Cut-set flows are defined by

$$f(X, \overline{X}) = \sum_{(i,j) \in (X, \overline{X})} f_{ij}, \tag{9}$$

$$f(\overline{X}, X) = \sum_{(j,i) \in (\overline{X}, X)} f_{ji}. \tag{10}$$

The conservation laws (6) and (7), when summed over $i \in X$, give

$$g = \sum_{i \in X} \left\{ \sum_{A(i)} f_{ij} - \sum_{B(i)} f_{ji} \right\},$$

that is,

$$g = f(X, \overline{X}) - f(\overline{X}, X), \tag{11}$$

since flows on links (i,j) with $i \in X$ and $j \in X$ cancel. Equation (11) is a mathematical statement of the theorem.[†]

It is important to emphasize that we must restrict our attention to cut-sets which separate the origin and destination. For example, in Fig. 8.1, node 1 is the origin, node 4 is the destination, and the flow value is $g = 7$. For $X = \{1, 2\}$, $\overline{X} = \{3, 4\}$,

---

[†] Equation (11) may be regarded as a discrete analog of the divergence theorem.

$$f(X, \overline{X}) - f(\overline{X}, X) = f_{13} + f_{23} + f_{24} - f_{32} = 4 + 1 + 5 - 3 = 7 = g.$$

But for $X = \{3\}$, $\overline{X} = \{1, 2, 4\}$,

$$f(X, \overline{X}) - f(\overline{X}, X) = f_{32} + f_{34} - f_{13} - f_{23} = 3 + 2 - 4 - 1 = 0 \neq g.$$

The net flow theorem has an important application in transportation studies. Suppose, for example, that an East–West river is used as a screen line and that traffic crossing the river is to be counted in order to calculate the traffic from a certain zone North of the river to a certain zone South of the river. Each motorist traveling South across the river is asked his origin and destination, and if these coincide with those being investigated his count is recorded. Does the total count necessarily give the correct O–D traffic? The answer is no, because the count only gives $f(X, \overline{X})$, and it is possible that a motorist on his trip may cross and recross the river. It is, therefore, essential to calculate the net flow by subtracting the quantity $f(\overline{X}, X)$, which represents the motorists travelling North across the river and yet going from the correct Northerly origin to the Southerly destination. In practice, of course, care is taken to choose a screen line so that $f(\overline{X}, X)$ can be neglected, but if the screen line is irregularly shaped and the origin and destination are close to it, some of the motorists' paths may in fact, cross and recross it.

The conservation equations (6)–(8) can be expressed in concise form by matrix notation. The *node–link incidence* matrix is an $n \times l$ matrix $\mathbf{E}$ whose element in the row corresponding to node i and the column corresponding to the link (j, k) is defined to be

$$+1, \quad \text{if} \quad i = j,$$
$$-1, \quad \text{if} \quad i = k,$$
$$0, \quad \text{otherwise.}$$

For the network illustrated in Fig. 8.1, the node–link incidence matrix is

$$
\mathbf{E} = \underset{i}{\text{node}}
\begin{array}{c}
 \\
1 \\
2 \\
3 \\
4
\end{array}
\overset{\displaystyle \text{link}\,(j,k)}{
\overset{(1,2)\ (1,3)\ (2,3)\ (2,4)\ (3,2)\ (3,4)}{
\left[
\begin{array}{cccccc}
1 & 1 & 0 & 0 & 0 & 0 \\
-1 & 0 & 1 & 1 & -1 & 0 \\
0 & -1 & -1 & 0 & 1 & 1 \\
0 & 0 & 0 & -1 & 0 & -1
\end{array}
\right]
}}. \qquad (12)
$$

Equations (6)–(8) can be written

$$\mathbf{Ef} = \mathbf{g}, \tag{13}$$

where $\mathbf{f}$ is the $l \times 1$ link flow vector and $\mathbf{g}$ is the $n \times 1$ O–D flow vector with first element $= g$, last element $= -g$, and all other elements $= 0$. We shall in the next chapter often refer to $\mathbf{f}$ as the *link flow pattern* or the *link flow traffic pattern*. For the example in Fig. 8.1,

$$\mathbf{f} = \begin{bmatrix} 3 \\ 4 \\ 1 \\ 5 \\ 3 \\ 2 \end{bmatrix}, \qquad \mathbf{g} = \begin{bmatrix} 7 \\ 0 \\ 0 \\ -7 \end{bmatrix}, \tag{14}$$

from which Eq. (13) can be verified with the use of Eq. (12).

The node–link incidence matrix has many interesting properties which are important in graph theory but are of not such interest in applications to transportation networks.

(c) Single O–D Network: Chain Flows

For transportation applications, an important example of flow on a single O–D network is obtained by the superposition of *chain flows* from the origin to the destination. In this formulation of network flow, it is convenient to denote the links of the network by $i = 1, 2, ..., l$; the link flows by f_i; the O–D chains by $m_j, j = 1, 2, ..., m$; and the chain flows by h_j. Then the flow value is given by the conservation equation

$$g = \sum_j h_j. \tag{15}$$

The link flows f_i resulting from the chain flows h_j can be obtained by letting

$$a_{ij} = \begin{cases} 1, & \text{if link } i \text{ is on chain } m_j, \\ 0, & \text{otherwise}, \end{cases} \tag{16}$$

so that

$$f_i = \sum_j a_{ij} h_j. \tag{17}$$

To obtain chain flows from given link flows is a rather different problem. For the example shown in Fig. 8.1, the link flows can be regarded as decomposed into the following chain-flows:

	Chain	Chain Flow
m_1:	$(1,2), (2,3), (3,4)$	1
m_2:	$(1,2), (2,4)$	2
m_3:	$(1,3), (3,2), (2,4)$	3
m_4:	$(1,3), (3,4)$	1

In this example the decomposition into chain flows is unique, but this is not generally true. It is possible that there are many different decompositions into chain flows, or it may be that no decomposition is possible. In order to illustrate the last point, it is only necessary to suppose that in Fig. 8.1 the link flows on $(2,3)$ and $(3,2)$ are increased to 101 and 103, respectively. One is then forced to consider the link flows as the superposition of chain flows given above plus a flow of 100 in the cycle $(2,3), (3,2)$. For our purposes, we shall neglect this possibility and always suppose that our single O–D network flows can be obtained by superposing chain flows. In our applications, we shall be interested in people and vehicles going from an origin to a destination and not those going aimlessly round in cycles!

This chain flow formulation can also be expressed in matrix form by introducing the *link–chain incidence matrix* $\mathbf{A}$ which is an $l \times m$ matrix with elements a_{ij} defined by Eq. (16). Then Eq. (17) can be written

$$\mathbf{f} = \mathbf{Ah}, \tag{18}$$

where $\mathbf{h}$ is the $m \times 1$ chain flow vector or *chain flow traffic pattern*. In addition, if we let $\mathbf{e}$ be the $m \times 1$ column with each element 1, and use a superscript T to denote matrix transposition, then Eq. (15) can be written

$$g = \mathbf{e}^{\mathrm{T}}\mathbf{h}. \tag{19}$$

For the example in Fig. 8.1,

$$
\mathbf{A} = \text{link} \quad
\begin{array}{c}
\\
(1,2) \\
(1,3) \\
(2,3) \\
(2,4) \\
(3,2) \\
(3,4)
\end{array}
\overset{\displaystyle \text{chain} \atop m_1\ m_2\ m_3\ m_4}{
\begin{bmatrix}
1 & 1 & 0 & 0 \\
0 & 0 & 1 & 1 \\
1 & 0 & 0 & 0 \\
0 & 1 & 1 & 0 \\
0 & 0 & 1 & 0 \\
1 & 0 & 0 & 1
\end{bmatrix}} \tag{20}
$$

and

$$
\mathbf{h} = \begin{bmatrix} 1 \\ 2 \\ 3 \\ 1 \end{bmatrix} \tag{21}
$$

from which Eq. (18) with Eq. (20) and (21) can be easily verified.

(d) Multiple O–D Network

It is simple to extend the notion of the single O–D network to networks with more than one O–D pair. The main difficulty is one of terminology and notation.

For transportation applications, it is essential to distinguish certain O–D flows from others, to make sure travelers get to their correct destinations. We can do this in either of two ways. First, we can consider the flow from each origin to all its destinations as a *copy*, each copy flow being distinguished by a superscript α. Secondly, we may consider each O–D flow from a particular origin to a particular destination as a flow of a separate *commodity*, denoted by superscript (k). For convenience, we shall use link flows for multicopy flow and chain flows for multicommodity flow.

For the multicopy flow description, let v^α be the copy flow from O^α and $v_j{}^\alpha$ the flow of this copy to destination node j. The extension of the link flow formulation for a single O–D network is immediate:

$$
\mathbf{E}\mathbf{f}^\alpha = \mathbf{g}^\alpha, \qquad \alpha = 1, 2, \ldots, \tag{22}
$$

where $\mathbf{E}$ is the $n \times l$ node–link incidence matrix, $\mathbf{f}^\alpha$ is the $l \times 1$ link flow vector with elements equal to the link flows of copy α, and $\mathbf{g}^\alpha$ the $n \times 1$ copy flow vector with elements

$$
g_i{}^\alpha = \begin{cases} v^\alpha, & \text{if } i \text{ is the origin of copy flow } \alpha, \\ -v_i{}^\alpha, & \text{if } i \text{ is a destination of copy flow } \alpha, \\ 0, & \text{otherwise.} \end{cases} \tag{23}
$$

The total link flow and the total network flow are given by superposing the copy flows:

$$\mathbf{f} = \sum_\alpha \mathbf{f}^\alpha, \tag{24}$$

$$v = \sum_\alpha v^\alpha = \sum_\alpha \sum_i v_i^\alpha. \tag{25}$$

Copy flows have been labeled according to their origin. It is evident that we could just as easily label them by their destinations so that a particular copy flow would be a flow from all origins to a particular destination.

For the alternative description of the multiple O–D network flow as a superposition of multicommodity flows, we denote all possible origin–destination pairs by $O^{(k)}$–$D^{(k)}$, $k = 1, 2, ..., q$, and let $m_j^{(k)}$ be the chains from $O^{(k)}$ to $D^{(k)}$. If $g^{(k)}$ is the flow of commodity k from $O^{(k)}$ to $D^{(k)}$, and $h_j^{(k)}$ the chain flow on $m_j^{(k)}$, then

$$g^{(k)} = \sum_j h_j^{(k)}. \tag{26}$$

If the links are denoted by $i = 1, 2, ..., l$, and

$$a_{ij}^{(k)} = \begin{cases} 1, & \text{if link } i \text{ is on chain } m_j^{(k)}, \\ 0, & \text{otherwise,} \end{cases} \tag{27}$$

then the link flow $f_i^{(k)}$ of flow from $O^{(k)}$ to $D^{(k)}$ is given by

$$f_i^{(k)} = \sum_j a_{ij}^{(k)} h_j^{(k)}. \tag{28}$$

The total network flow is

$$v = \sum_k g^{(k)}, \tag{29}$$

and the total link flow f_i on link i is

$$f_i = \sum_k f_i^{(k)} = \sum_k \sum_j a_{ij}^{(k)} h_j^{(k)}. \tag{30}$$

Equations (26)–(28) are obvious generalizations of (15)–(17).

Equations (26) and (28) can be written in matrix notation, which is an obvious extension of that used above. Thus, (26) becomes

$$g^{(k)} = (\mathbf{e}^{(k)})^{\mathrm{T}} \mathbf{h}^{(k)} \tag{31}$$

where $\mathbf{e}^{(k)}$ is a column with all entries equal to unity, and (28) becomes

$$\mathbf{f}^{(k)} = \mathbf{A}^{(k)} \mathbf{h}^{(k)}, \tag{32}$$

where $\mathbf{A}^{(k)}$ is the link–chain matrix.

Although (30) may appear unduly complicated, it explicitly reflects what a traveler on a particular stretch of road would well realize. In the traffic stream there are:

(i)   travelers with the same origin and destination who have chosen his precise route (same values of i, j, k);

(ii)   travelers with the same origin and destination whose routes are different but happen to coincide with his own on this particular road segment (same values of i and k, different j for which $a_{ij}^{(k)} = 1$);

(iii)   travelers with different origins and/or destinations whose routes happen to coincide with his own on this particular road segment (same i, different k for which $a_{ij}^{(k)} = 1$).

The distinction we have made between multicopy and multicommodity flows is a compromise between varied uses of the terms in the literature. For our purposes, especially in applications to traffic assignment, the terminology and notation we have adopted proves most convenient. In order to distinguish flow on a tree with one home node from a tree with another home node, we use multicopy flows. When the distinction is from flow on one chain to flow on another, we use multicommodity flows.

(e) Compressibility and Separability

In some of the transportation network problems that we discuss in this text, other conservation equations besides those derived from Kirchhoff's law will be imposed. It frequently happens that the graph $[N; L]$ of a given transportation network is *compressed* or *expanded* into a new network $[N'; L']$. The compressing process may occur, for example, when several nodes and links in $[N; L]$ are combined to form $[N'; L']$; hence an analysis of flows on the larger network $[N; L]$ is replaced by one on the smaller network $[N'; L']$, where $N' \subset N, L' \subset L$. Of course, Kirchhoff's law must hold on both $[N; L]$ and $[N'; L']$, but we also impose the *compressibility* requirement that

$$f'_{ij} = \sum f_{ks}, \tag{33}$$

where the summation extends over the links (k, s) in L that are combined to form the link (i, j) in L'.

Consider the numerical example illustrated in Fig. 8.2(a), where

$$N = \{1, 2, 3, 4\}$$

$$L = \{(1,2), (1,3), (1,4), (2,3), (2,4), (3,2), (3,3), (3,4)\},$$

and the flows in $[N; L]$ are given by

$$f = \{f_{12}, f_{13}, f_{14}, f_{23}, f_{24}, f_{32}, f_{33}, f_{34}\}$$

$$= \{1, 2, 3, 2, 2, 4, 1, 1\}.$$

For purposes of this illustration we combine nodes 2 and 3 to obtain the network in Fig. 8.3, with $N' = \{1, 2, 3\}$, $L' = \{(1,2), (2,2), (2,3), (1,3)\}$.

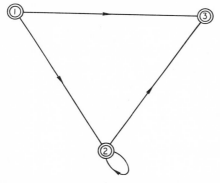

Figure 8.3.   Compressed network obtained by combining nodes 2 and 3 of the network in Fig. 8.2(a).

node	i	1	2	3
production	a_i	6	10	0
attraction	b_i	0	10	6

Notice that node 2 in N' corresponds to nodes 2 and 3 in N, node 3 in N' to node 4 in N, i.e., there is a renumbering of nodes. Similarly links $(1,2)$, $(1,3)$ in L are replaced by $(1,2)$ in L', links $(2,4)$ and $(3,4)$ in L by $(2,3)$ in L' and links $(2,3)$, $(3,2)$, and $(3,3)$ in L by $(2,2)$ in L'. The compressibility equations are

$$f'_{12} = f_{12} + f_{13} \qquad = 3, \tag{34}$$

$$f'_{13} = f_{14} \qquad = 3, \tag{35}$$

$$f'_{22} = f_{23} + f_{32} + f_{33} = 7, \tag{36}$$

$$f'_{23} = f_{24} + f_{34} \qquad = 3. \tag{37}$$

It will be noted that the production and attraction at the new node 2 are the sums of the productions and attractions for the original nodes 2, 3. It is also evident that Kirchhoff's law holds in the compressed network.

If the network is *expanded* to give a more detailed network with more nodes and links, the compressibility equations must hold in the reverse direction.

Another obvious requirement which the network flows must satisfy is *separability*, in the sense that, if a subgraph is obtained by removing one node and its attached links, the flows on the untouched links must remain unaltered. For example, if node 4 in Fig. 8.2(a) is removed, the new network in Fig. 8.4 is obtained. Notice that the production and attractions are adjusted to compensate for the excluded links.

In later chapters, where mathematical models of flow in transportation networks are described, it will be shown that the requirements of compressibility and separability are often desirable model properties.

9.   Costs and Capacities

(a) Link, Route, and Network Costs

Up to this point, in our outline of the elements of network theory, we have considered the structure of graphs and transportation networks and the natural requirements of flow conservation. It is characteristic of almost all transportation problems we study that the passage of flow through the network creates delays, incurs costs, or in some way affects the behavior of traffic movement. For example, one may associate with a link of a network the average travel time in traversing a street segment, the distance between two intersections, the payment of money for the transport of goods along a road, or a toll levied on users of a freeway. In order to obtain a certain degree of uniformity in notation, we shall refer to link, route, and network *costs* with the understanding that, depending on the particular usage, costs may refer to money, travel times, delays, distances, disutility, or perhaps combinations of these.

We introduce the notion of the *link cost* $c_{ij}(f_{ij})$ as an average cost or cost per unit flow by defining the *total link cost* for a link (i, j) with flow f_{ij} as

$$f_{ij} c_{ij}(f_{ij}).$$

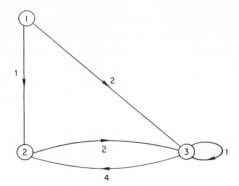

Figure 8.4. Separated network obtained from Fig. 8.2(a) by removing and excluding node 4.

node	i	1	2	3
production	a_i	3	2	5
attraction	b_i	0	5	5

The notation indicates that the link cost is, in general, a function of the link flow and that this function may differ for different links. Furthermore, all flow units in link (i,j) perceive the same link cost. If we use the notation i for a link, the link cost is denoted by $c_i(f_i)$. In the special and important case when the link cost is *flow independent*, we use the notations c_{ij}, $c(n_i,n_j)$, c_i, or $c(l_i)$ for the links denoted by (i,j), (n_i,n_j), i, or l_i.

Next we define the *route cost* as the cost of unit flow on a route, such as a chain or path, from an origin to a destination of a network. In general, the link costs will be additive, so that the route cost is the sum of the link costs together, perhaps, with the costs of traversing the nodes. These latter costs may be intersection delays or penalties for turns. For a main road network the detailed traffic movements at individual intersections are unimportant, and the costs of traversing nodes can sometimes be included in the link costs. For example, the travel time on a link may be conceived as the time for travel from the beginning of the link to the beginning of the next link. If the costs for traversing nodes do not have to be included explicitly, the route cost of unit chain flow on the chain, denoted by the node sequence $n_1, n_2, ..., n_r$, from an origin n_1 to a destination n_r, is given by

$$C(n_1, n_r) = \sum_{i=1}^{r-1} c(n_i, n_{i+1}),$$ (1)

where, for simplicity, flow-independent link costs have been assumed.

For city street networks and other networks where intersection delays are important, a more general definition of route cost is required. Suppose that the flow-independent link cost associated with link l_i is denoted by $c(l_i)$, and the penalty associated with a turn from link l_i to link l_j by $p(l_i, l_j)$. Without loss of generality and with some advantage, the origin of a route from l_1 to l_r can be interpreted as the beginning of the link l_1 (on the far side of the intersection represented by the node from which l_1 is directed) and likewise the destination is taken as the beginning of the link l_r. The route cost of unit chain flow on the chain denoted by the link sequence $l_1, l_2, ..., l_r$ is then defined to be

$$C(l_1, l_r) = \sum_{i=1}^{r-1} [c(l_i) + p(l_i, l_{i+1})]. \tag{2}$$

It might well be asked why, in defining route costs, sequences of nodes should be used in Definition (1) and sequences of links in Definition (2). The reason for this will become evident when the cheapest route problem is discussed in Sect. 14, Chap. III.

Finally, we define the *network cost* C as the sum of the total link costs for all links of the network, i.e.,

$$C = \sum_{(i,j)} f_{ij} c_{ij}(f_{ij}). \tag{3}$$

It is worth noting that a prohibited link or turn can be represented by an infinite cost. Thus if flow on the link (i,j) is prohibited, we can take $c_{ij} = \infty$, and likewise if the turn from link l_i to l_j is prohibited, we can take $p(l_i, l_j) = \infty$.

For a single O–D network (without turn penalties), the link flows can be regarded as the superposition of chain flows $h_j, j = 1, 2, ..., m$ on the O–D chains m_j. For flow-dependent link cost, the route cost C_j, on the jth chain is, in general, a function of all chain flows $h_1, h_2, ..., h_m$, since different chains can share the same links. We therefore write

$$C_j(h_1, h_2, ..., h_m) = C_j(\mathbf{h}), \tag{4}$$

for the route cost and

$$C = \sum_j h_j C_j(\mathbf{h}) = \mathbf{C}^{\mathrm{T}} \mathbf{h}, \tag{5}$$

for the network cost. The extension of these definitions to multiple O–D networks is straightforward.

(b) Capacitated Network

In the discussion so far, there have been no upper bounds on the link flows—they could be any nonnegative values provided they satisfy the conservation laws. For applications to transportation networks, it is sometimes important to consider a *capacitated network* in which link flows must obey the inequalities

$$0 \leqslant f_{ij} \leqslant u_{ij}. \tag{6}$$

The quantities u_{ij} are the *link capacities* and with little loss in generality can be assumed to be positive integers (see Fig. 9.1).

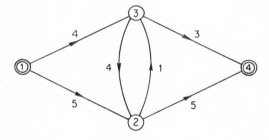

Figure 9.1. A single O–D capacitated network with link capacities as indicated. The link flows given in Fig. 8.1 give a feasible flow for the above network.

In designing a future network, the traffic engineer often uses the celebrated Highway Capacity Manual [7] as a practical guide in helping to determine the capacities of streets and intersections as a function of road widths, number of lanes, shoulder widths, gradients, traffic signalization, etc. Depending on the level of service to be provided, capacities are chosen and the future network tested by checking whether the estimated traffic exceeds any capacities. If so, more capacity can be provided by designing improved facilities. It is clear, then, that inequalities such as (6) are significant for the planning process.

An interesting point worth noting is that it is the intersections rather than the streets which are potential bottlenecks in a city street network. The emphasis on link capacities in flow theory is more appropriate to a main road network where the nodes are not of direct traffic significance. But it is not difficult to include node capacities—they are easily introduced by representing a capacitated node by two nodes joined by one capacitated dummy link. Partly for this reason, it is usual to suppose that only the links of a network are capacitated.

A set of link flows satisfying the conservation equations and the constraint Eq. (6) is called a *feasible network flow*. For example, the link flows in Fig. 8.1 give a feasible network flow for the capacitated network in Fig. 9.1. In matrix notation, the constraints on link flow for a single O–D capacitated network may be written

$$0 \leqslant \mathbf{f} \leqslant \mathbf{u} \tag{7}$$

$$\mathbf{Ef} = \mathbf{g}. \tag{8}$$

For a capacitated network, the *cut capacity* associated with a cut-set $(X, \overline{X})$ is defined by

$$u(X, \overline{X}) = \sum_{(i,j) \in (X, \overline{X})} u_{ij}. \tag{9}$$

For the network in Fig. 9.1, the cut-capacities for cut-sets separating the origin from the destination are as follows:

X	$u(X, \overline{X})$
$\{1\}$	9
$\{1, 2\}$	10
$\{1, 2, 3\}$	8
$\{1, 3\}$	12

The cut capacities give an upper bound to the flow through a network. Consider, for example, a single O–D capacitated network, and let $(X, \overline{X})$ be any cut-set separating the origin from the destination. Then it is a simple consequence of the net flow theorem proved in Sect. 8 that

$$g \leqslant u(X, \overline{X}). \tag{10}$$

This follows from

$$f_{ij} \leqslant u_{ij} \Rightarrow f(X, \overline{X}) \leqslant u(X, \overline{X}),$$

$$f_{ij} \geqslant 0 \Rightarrow f(\overline{X}, X) \geqslant 0,$$

which together give

$$g = f(X, \overline{X}) - f(\overline{X}, X) \leqslant u(X, \overline{X}).$$

For example, the flow value $g = 7$ for the network flows in Fig. 8.1 is less than the cut capacities listed above.

The relation (10) has an interesting interpretation when applied to screen-lines for main road networks. Suppose, for example, that an East–West river is used as a screen line for the study area of a transportation analysis. Then (10) implies that the total flow of traffic from origins North of the river to destinations South of the river is limited by the total capacities of all Southbound roads or bridges crossing the river.

One of the central results of network theory is embodied in the max-flow min-cut theorem (see [1] and also Sect. 18, Chap. III, Problem 18) which goes further than (10) and states that the maximal flow g^* (i.e., the maximum flow value for all feasible flows) is equal to the minimum cut capacity, that is,

$$g^* = \min_X u(X, \bar{X}). \tag{11}$$

Thus, the maximal flow for the capacitated network in Fig. 9.1 is 8 units of flow. It is left as an exercise for the reader to see how to increase the flow value of 7 in Fig. 8.1 to a maximal flow of 8.

The identification of a cut-set with minimum cut capacity is often obvious for a transportation network, e.g., for a study area divided by a river or other obstacle with few crossings. On the other hand, it often occurs that congested traffic conditions arise long before the theoretical maximal flows are achieved. This is partly because of the effects of intersections in delaying and platooning traffic, and partly because travelers tend to ignore minor roads although they may be operating well below capacity.

To conclude this section, it is worth mentioning that link capacities could be implied in the concept of link costs. For example,

$$c_{ij}(f_{ij}) = \infty, \quad \text{for} \quad f_{ij} < 0, \quad \text{or} \quad f_{ij} > u_{ij}, \tag{12}$$

effectively implies (6).

10. Conclusion

In this chapter, we have briefly summarized the concepts and notations of network theory which we shall use in discussing problems of flow on transportation networks. The choice of terminology and symbols has required a considerable compromise among the variety in use in the literature. Where possible we have opted for the usage most common in transportation applications.

11. Notes and References

There is now an extensive literature on graph theory and network flows and the following texts are all excellent references:

[1]   Ford, L. R. and Fulkerson, D. R., *Flows in Networks*, Princeton Univ. Press, Princeton, New Jersey (1962).

This book may be regarded as the bible on the general subject of network flows, and contains careful proofs of fundamental theorems. It covers a wide range of applications, but they are rather different from the scope of this text.

[2]   Berge, C. and Ghouila-Houri, A., *Programming, Games and Transportation Networks*, Wiley, New York (1965).

This is a comprehensive text, mathematical in approach. The reader is warned that the authors' definition of a path and chain is opposite to that used by Ford and Fulkerson (and adopted in this text). Unfortunately, this is indicative of the extreme confusion in the literature in the choice of terminology and notation for graph theoretical concepts.

[3]   Busacker, R. G. and Saaty, T. L., *Finite Graphs and Applications*, McGraw-Hill, New York (1965).

This is an extremely readable text, particularly the chapter correctly titled "A Variety of Interesting Applications." Those who have been frustrated by the widely marketed "Instant Insanity" puzzle will find in this chapter an elegant solution using graph theory techniques.

[4]   Kaufmann, A., *Graphs, Dynamic Programming and Finite Games*, Academic Press, New York (1967).

The exposition in this text is noteworthy for the examples and industrial applications which the author uses to illustrate general graph theory concepts.

[5]   Harary, F. (ed.), *Graph Theory and Theoretical Physics*, Academic Press, New York (1967).

This book contains an interesting series of articles based on presentations at a NATO Summer School. The opening section of the first chapter discusses some of the terminology used in graph theory "with a display of some of the chaos which runs rampant in this field." The chapter on electrical networks makes an interesting comparison with the present text.

[6] Beckmann, M., McGuire, C. B. and Winsten, C. B., *Studies in the Economics of Transportation*, Yale Univ. Press, New Haven, Connecticut (1956).

This excellent book has greatly influenced the development of transportation science. The first part is devoted to an analytical study of highway transportation and several chapters are immediately relevant to this text. Although the book was published some years ago, it still reads as an up-to-date survey of research on transportation problems.

A practical guide to the determination of highway capacities is given in the following:

[7] Highway Research Board, *Highway Capacity Manual*, Special Report 87 (1965).

12.   Problems

The following problems, except for the last, all relate to the directed network defined by

$N = \{1, 2, 3, 4, 5, 6\}$,

$L = \{(1, 2), (1, 3), (2, 3), (2, 4), (2, 5), (3, 4), (3, 5), (4, 5), (4, 6), (5, 6)\}$.

1. Is the graph complete?

2. Is the graph connected?

3. List all chains from node 1 to node 6.

4. List all paths from node 1 to node 6 which are not chains.

5. List all meshes in the graph which do not include nodes 1 and 6.

6. Give two examples of spanning trees of the graph, one of which is and the other is not an arborescence with node 1 as home node.

7. Write down the node–link incidence matrix E.

8. Write down the link–chain incidence matrix A for chains from node 1 to node 6.

9. Compute the matrix product **EA**.

10. If node 1 is an origin of flow and node 6 a destination, and the link flows are

$$f_{12} = 5, \quad f_{13} = 7, \quad f_{23} = 0, \quad f_{24} = 4, \quad f_{25} = 1,$$
$$f_{34} = 4, \quad f_{35} = 3, \quad f_{45} = 4, \quad f_{46} = 4, \quad f_{56} = 8,$$

find the flow value g, check Kirchhoff's law for nodes 5 and 6, and verify that

$$\textbf{Ef} = \textbf{g}.$$

11. Find a set of chain flows giving the link flows listed in Problem 10 and verify that

$$\textbf{f} = \textbf{Ah}.$$

Is the set of chain flows unique?

12. For the link flows listed in Problem 10, verify the net flow theorem for the cut-set with $X = \{1, 2, 5\}$.

13. If the link costs are

$$c_{12} = 1, \quad c_{13} = 4, \quad c_{23} = 2, \quad c_{24} = 6, \quad c_{25} = 8,$$
$$c_{34} = 3, \quad c_{35} = 3, \quad c_{45} = 1, \quad c_{46} = 1, \quad c_{56} = 1,$$

find:

(a) the route cost of the chain from node 1 to node 6 which has the greatest number of links;
(b) the chain from node 1 to node 6 which has the minimum route cost;
(c) the network cost for the link flows listed in Problem 10.

14. If the link capacities are given by

$$u_{12} = 6, \quad u_{13} = 8, \quad u_{23} = 2, \quad u_{24} = 4, \quad u_{25} = 1,$$
$$u_{34} = 5, \quad u_{35} = 3, \quad u_{45} = 4, \quad u_{46} = 8, \quad u_{56} = 9,$$

find the cut capacities for all cut-sets separating node 1 from node 6.

15. For the link capacities given in Problem 14, is the network flow given by the link flows in Problem 10 (a) feasible (b) maximal?

16. This problem refers to the street network illustrated in Fig. 2.1. Suppose that the capacity of Battery, Clay, and Kearny Streets is 60 units, that of Pacific is 50 units each way, that of all other one-way streets is 20 units and that of all other two-way streets 10 units each way.

(a) Find the maximal traffic flow from node 1 to node 52 and a cut-set with minimum cut capacity. (Note that, by (10), Sect. 9, it is sufficient to find chain flows and a cut-set such that the flow value is equal to the cut capacity.)
(b) Repeat the calculation if the road segments (10, 11) and (35, 36) are closed to traffic.

III

EXTREMAL PRINCIPLES AND TRAFFIC ASSIGNMENT

13. Introduction

The elements of network theory as outlined in the previous chapter will serve as a foundation for our analysis of traffic movement in transportation networks. As pointed out in Sect. 3, Chap. I, the traffic flow on road networks exhibits certain regularities and patterns which the mathematical models used in the transportation process attempt to describe. In surmising whether this regularity of traffic movement might be formulated in terms of general laws or principles, it is useful to consider, by analogy, the historical development of gravitational theory. By careful observation of the movements of the planets, Kepler detected certain regularities, from which Newton, with his mathematical knowledge, was able to infer the inverse square law and the momentum principles. One cannot expect that traffic movement could be described by such simple, powerful, and all-embracing principles as Newton's laws, especially because of the influence of the human being as a decision maker in the driver's seat! Nevertheless, one can expect that there are some broad principles which can serve as a useful basis for modeling traffic flow—and it has to be remembered that in planning future roads, the accuracy needed to pinpoint a moon landing is not required.

It is the purpose of this chapter to investigate the implications of two broad principles which were first enunciated by Wardrop [1]. He gave

49

the following two criteria for determining the distribution of traffic over alternative routes:

(i)   *"The journey time on all routes actually used are equal, and less than those which would be experienced by a single vehicle on any unused route."*

(ii) *"The average journey time is a minimum."*

Wardrop compared these two as follows: "The first criterion is quite a likely one in practice, since it might be assumed that traffic will tend to settle down into an equilibrium situation in which no driver can reduce his journey time by choosing a new route. On the other hand, the second criterion is the most efficient in the sense that it minimizes the vehicle-hours spent on the journey."

In anticipation of a description to be elaborated later, we shall call traffic patterns which are optimized according to the first criterion *user-optimized*, and *system-optimized* patterns those which are optimal according to the second principle.

The formulation of these two criteria, or *extremal principles* as we shall call them, depends on the properties of the transportation network being described. In our terminology of cost rather than travel time the application of the extremal principles involves two important network flow problems: first, the determination of cheapest routes on a network; and second, the minimization of total network cost.

The next section of this chapter will be devoted to a discussion of cheapest route algorithms with special emphasis on those which have been used in transportation planning program packages.

The succeeding section will analyze the minimum network cost problem for a single O–D network with link costs which are dependent on flow. The system-optimized traffic patterns will be compared and contrasted with the cheapest route patterns and their relation described by what we call the *principle of available chains*.

Finally, we consider the more complicated, but more significant, case of a multiple O–D network with costs which are dependent on flow. By means of two theorems, we shall demonstrate how user-optimized and system-optimized patterns differ and yet are intimately related.

The extremal principles have formed the basis of a variety of traffic assignment models which have been used in transportation planning for the allocation of traffic to road networks. Where appropriate, we describe some of these models and their relation to the extremal principles.

14.   Cheapest Routes

(a) Appraisal of Algorithms

There is an extensive literature on the problem of determining the cheapest routes on a network; the bibliography prepared by Murchland [2] lists about one hundred references. Many algorithms have been proposed, some have subsequently been shown to be incorrect, others have proved inferior to earlier ones, and indeed the literature presents a very confusing picture. This has been greatly clarified by a recent article by Dreyfus [3] which gives an appraisal of algorithms proposed for networks without turn penalties. A similar article [4] considers algorithms for networks with turn penalties, but a discussion of these will be deferred until Sect. 14(c).

For networks without turn penalties, the route cost as defined in Sect. 9(a), Chap. II, is the sum of the link costs, and in our terminology the routes themselves are chains (but not paths) if the network is directed, and chains or paths if the network is undirected. The problem of determining the cheapest routes (variously called shortest paths, quickest paths, minimum routes, etc.) can be stated in several forms, depending on whether one requires the cheapest routes between two specified nodes, between a specified node and all other nodes, or between all pairs of nodes. A further complication arises if negative link costs are permitted, but for transportation applications this possibility can be ignored.

All the acceptable algorithms are in essence based on the following property: if a cheapest route from node n_1 to node n_r passes through node n_i, then that portion of the route from n_1 to n_i is a cheapest route from n_1 to n_i (and likewise the portion of the route from n_i to n_r is a cheapest route from n_i to n_r). As Bellman [5] has shown, this fundamental property allows the cheapest route problem to be expressed as a dynamic program embodied in the following

THEOREM: *The costs of the cheapest routes from node n_1 to nodes n_r of a network $[N; L]$ with positive link costs $c(n_i, n_j)$ are the unique solutions of the functional equations*

$$f(n_1) = 0, \tag{1}$$

$$f(n_r) = \min_{n_i \neq n_r} [f(n_i) + c(n_i, n_r)], \qquad n_r \neq n_1. \tag{2}$$

A proof of this theorem is given in Appendix A, and even a cursory glance at the derivation should convince the reader that the verification of a network algorithm is not a trivial matter. Precise theorems and proofs are required and, as the literature shows, incorrect algorithms seem plausible since their faults are often not obvious until counterexamples are exhibited.

The various correct cheapest route algorithms are in effect different procedures for solving the functional equations (1) and (2). The efficiency of the different algorithms can be rated theoretically by comparing the number of iterations, additions, and comparisons they require. But this may not be a helpful guide to the computing efficiency one can expect when the algorithms are applied to transportation networks, whose special structure can often be exploited with dramatic savings in computer time and storage.

In the transportation planning process it is necessary to determine the cheapest routes between centroids of main road networks. This is commonly achieved by using a tree-building algorithm which builds successively, for each centroid as home node, a cheapest route-spanning tree. Many variants of the tree-building algorithm are in use, and one is described in general terms in the following.

(b) Tree-Building Algorithms

The tree-building algorithms used in transportation planning programs solve the functional equations (1) and (2) by fanning out from the home node to all other nodes in increasing order of their costs from the home node [6]. The algorithm will first be described in words, then formulated algebraically, and finally applied to a simple example.

The nodes are successively labeled with two numbers, one the predecessor node on a cheapest route from the home node and the other the cost of the route. Initially, the home node is permanently labeled $(0, 0)$ and all other nodes are tentatively labelled $(0, \infty)$. The general step is in two parts. The last node permanently labeled is scanned in the following way: for all nodes to which links from this node are directed the following comparison is made. Is the sum of the cost label on the node being scanned and the link cost less than the tentative node label? If the answer is yes, the node being scanned becomes the new tentative predecessor node label and the lesser cost the new tentative cost label. All other tentative labels are left unchanged. The second part of the general step is to compare the tentative cost labels and declare a node with the minimum such label permanently labeled. The general step is then repeated until the permanent labeling is completed.[†]

This procedure is simple to follow through for a particular example but is a little complicated to formulate algebraically. It is convenient to duplicate the notation by denoting the nodes $1, 2, ..., i, ...$ and also $n_1, n_2, ..., n_k, ...$ as they are successively selected at steps $1, 2, ..., k, ...$ of the algorithm. For simplicity, it will be supposed that the network $[N; L]$ is connected and undirected and that the link costs are $c(n_i, n_j) > 0$. At the kth step, nodes n_j will be tentatively labeled with two numbers, $P^{(k)}(n_j)$, the current predecessor node, and $C^{(k)}(n_1, n_j)$, the cost of the currently known cheapest path from the home node n_1 to n_j. At the conclusion of the algorithm, all nodes have permanent labels $P^*(n_j)$ and $C^*(n_1, n_j)$ representing the predecessor node and the cost of a cheapest path, respectively. The algorithm can be formulated as follows:

Step $k = 1$:

$$X_1 = \{n_1\}, \tag{3}$$

$$P^*(n_1) = 0, \qquad C^*(n_1, n_1) = 0, \tag{4}$$

$$P^{(1)}(n_j) = 0, \qquad C^{(1)}(n_1, n_j) = \infty, \qquad n_j \neq n_1. \tag{5}$$

Steps $k = 2, 3, ..., n$:

(i) For

$$n_j \in \bar{X}_{k-1}, \qquad (n_{k-1}, n_j) \in (X_{k-1}, \bar{X}_{k-1}),$$

$$C^*(n_1, n_{k-1}) + c(n_{k-1}, n_j) < C^{(k-1)}(n_1, n_j),$$

define

$$P^{(k)}(n_j) = n_{k-1}, \tag{6}$$

and

$$C^{(k)}(n_1, n_j) = C^*(n_1, n_{k-1}) + c(n_{k-1}, n_j). \tag{7}$$

(ii) For all other $n_j \in \bar{X}_{k-1}$, define

$$P^{(k)}(n_j) = P^{(k-1)}(n_j), \tag{8}$$

[†] It is interesting to note that when all link costs are assigned the value 1, this algorithm finds which nodes are accessible from a given origin node.

TABLE 14.1   Tree-Building Algorithm for Fig. 14.1, Home Node 1[a]

Step k	Nodes i	Tentative labels Predecessor $P^{(k)}$	Cost $C^{(k)}$	Node n_k	Permanent labels Predecessor P^*	Cost C^*
1	1	0	0	1	0	0
	2–24	0	∞			
2	10	1	5	10	1	5
	11	1	12			
	12	1	13			
	2–9, 13–24	0	∞			
3	24	10	15			
	11	1	12	11	1	12
	12	1	13			
	2–9, 13–23	0	∞			
4	20	11	27			
	24	10	15			
	12	1	13	12	1	13
	2–9, 13–19, 21–23	0	∞			
5	13	12	21			
	20	11	27			
	24	10	15	24	10	15
	2–9, 14–19, 21–23	0	∞			
6	9	24	23			
	22	24	34			
	23	24	40			
	13	12	21	13	12	21
	20	11	27			
	2–8, 14–19, 21	0	∞			
7–20						
21	16	15	95			
	17	18	60	17	18	60
	4, 5	0	∞			
22	16	17	70			
	5	17	65	5	17	65
	4	0	∞			
23	16	17	70	16	17	70
	4	0	∞			
24	4	16	75	4	16	75

[a] Details of Steps $k = 7$–20 are omitted.

and

$$C^{(k)}(n_1, n_j) = C^{(k-1)}(n_1, n_j). \tag{9}$$

(iii) Define n_k by

$$C^{(k)}(n_1, n_k) = \min_{n_j \in \bar{X}_{k-1}} C^{(k)}(n_1, n_j), \tag{10}$$

and define the permanent labels

$$P^*(n_k) = P^{(k)}(n_k), \tag{11}$$

$$C^*(n_1, n_k) = C^{(k)}(n_1, n_k). \tag{12}$$

(iv) Define

$$X_k = X_{k-1} \cup n_k. \tag{13}$$

Since $C^*(n_1, n_k) \geqslant C^*(n_1, n_{k-1})$, this procedure solves (1) and (2) by permanently labeling nodes in increasing order of cost from the home node. The cheapest paths themselves can be determined by tracing back to the home node *via* successive predecessor nodes. For any link $(P^*(n_k), n_k)$ on a cheapest path, (7) implies that

$$C^*(n_1, n_k) - C^*(n_1, P^*(n_k)) = c(P^*(n_k), n_k), \tag{14}$$

and if

$$C^*(n_1, n_j) - C^*(n_1, n_i) < c(n_i, n_j), \tag{15}$$

then (n_i, n_j) is not on a cheapest path.

This tree-building algorithm will be illustrated by computing a cheapest spanning tree for the undirected network illustrated in Fig. 14.1. The first 6 and the last 4 steps in the algorithm for home node 1 are given in Table 14.1, the complete "tree trace" in Table 14.2, and the tree itself is illustrated in Fig. 14.2. Because the network is undirected, the tree also represents the cheapest paths from all nodes to the home node. For an example of this size, it is convenient and simple to label the nodes in pencil, erase the labels when necessary (e.g., as in Step 21), keep an updated list of tentative cost labels, and indicate with an asterisk or tick when a label is declared permanent.

As input to trip distribution models (to be discussed in Chap. IV), the costs of cheapest routes between centroids may be required, and these can be obtained by a "skim tree" procedure which selects the required costs from the output of the cheapest route algorithm applied for each

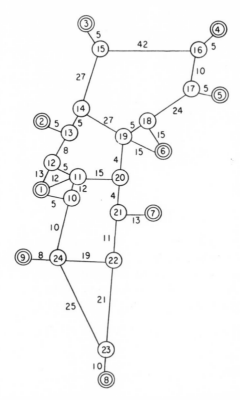

Figure 14.1.   Main road network for the Bay Area. The network is the same as that illustrated in Fig. 2.4, except that the dummy links have not been distinguished from the other links. The indicated link costs are approximate travel times in minutes.

centroid of the main road network as home node. In the present example, the tree trace for home node 1 is skimmed by reporting the values of C^* for the eight other centroids $i = 2, 3, ..., 9$.

(c) Turn Penalties and Prohibitions

As indicated in Sect. 9(a), Chap. II, it is important in analyzing city street networks and other transportation networks where delays at intersections are significant to include, in addition to link costs, penalties for turns at nodes. A turn prohibition can be regarded as a turn with infinite penalty.

The various procedures which have been suggested for determining cheapest routes for networks with turn penalties and prohibitions are appraised in [4]. The first relevant publication is the short paper by Caldwell [7] which shows that turn penalties can be theoretically taken into account by constructing a pseudo-network in which nodes represent the original links and links represent "hooks" or ordered pairs of links.

TABLE 14.2 CHEAPEST TREE
TRACE FOR HOME NODE 1

Node i	Predecessor $P*$	Cost $C*$
1	0	0
2	13	26
3	15	58
4	16	75
5	17	65
6	19	46
7	21	44
8	23	50
9	24	23
10	1	5
11	1	12
12	1	13
13	12	21
14	13	26
15	14	53
16	17	70
17	18	60
18	19	36
19	20	31
20	11	27
21	20	31
22	24	34
23	24	40
24	10	15

An important sentence in the paper is worth quoting: "It isn't necessarily true that the best path from the origin to a node i through a node j coincides, from the origin to j, with the best path from the origin to j." The network shown in Fig. 14.3(a) illustrates this. The numbers ascribed

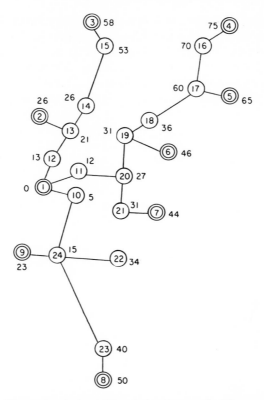

Figure 14.2   Cheapest route tree with node 1 as home node. The numbers attached to the nodes are the costs of the cheapest paths from the home node.

to the links are interpreted as link costs and each turn is assumed penalized with a cost of 2 units. The following route costs:

route	cost
1, 2, 4	$5 + 2 + 3 = 10$
1, 3, 4	$4 + 2 + 5 = 11$
1, 3, 4, 5	$4 + 2 + 5 + 5 = 16$
1, 2, 4, 5	$5 + 2 + 3 + 2 + 5 = 17$

illustrate that 1, 2, 4 is the cheapest route from node 1 to node 4, whereas 1, 3, 4, 5 is the cheapest route from node 1 to node 5.

The standard tree-building programs in transportation planning computer packages [8,9] attempt to account for turn penalties by labeling each North–South link with a "plus" sign and each East–West link with a "minus" sign, as illustrated in Fig. 14.3(b). As the cheapest route trees are built, a change in sign in passing from one link to another is used to indicate that a certain constant turn penalty should be added. Although this heuristic procedure tends to eliminate the building of "illogical" stair-case routes and is adequate for many networks, it has become well known by users of the programs that it is not strictly correct. For the example in Fig. 14.3(a), the algorithm would correctly obtain route 1, 2, 4 as the cheapest route from node 1 to node 4 but having discarded the alternative route 1, 3, 4 would incorrectly obtain 1, 2, 4, 5 as the cheapest route from node 1 to node 5. The method has been modified so that a variety of turn penalties can be included in specified intersection types, but the basic invalidity of the procedure is, of course, not overcome.

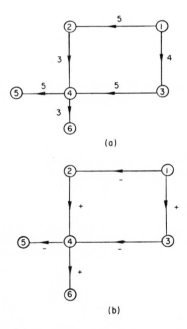

(a)

(b)

Figure 14.3. Networks with turn penalties. In addition to the link costs indicated in (a), each turn is assumed to have a penalty of 2 units. Some computer programs attempt to take these into account by using plus signs and minus signs as in (b) and penalizing each change in sign.

The tree-building programs in the same computer packages attempt to allow for turn prohibitions by listing them separately and excluding a route when a prohibition is indicated. Again, it is known that this heuristic procedure is not valid and the simple example in Fig. 14.4(a)

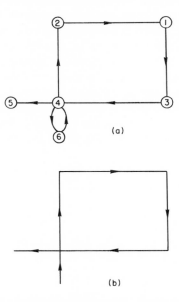

Figure 14.4.   Network with prohibited turn. (a) The left turn from link (6, 4) to link (4, 5) is assumed prohibited. There is no chain from node 6 to node 5, but there is an acceptable route 6, 4, 2, 1, 3, 4, 5 which visits node 4 twice. (b) Traffic sign indicating how a prohibited left turn should be negotiated.

illustrates the difficulties its use can lead to. The algorithm, like all those for networks without turn penalties and prohibitions, considers only those routes which are chains (or paths if the network is undirected), strictly defined as sequences of *distinct* nodes with consecutive nodes joined by links. If, in Fig. 14.4(a), the left turn from link (6, 4) to link (4, 5) is prohibited, there is no chain from node 6 to node 5, whereas the route *via* nodes 6, 4, 2, 1, 3, 4, 5 would be an acceptable or logical route. The importance of including such routes in transportation networks is exemplified by the use of signs in the heart of West London, England, which indicate to a motorist how he should negotiate a prohibited turn from an arterial road into a minor cross street. The signs, displayed on the approaches to the intersections, are a keep-to-the-left version of Fig. 14.4(b). The significance of Caldwell's pseudo-network in regard to this problem, although not discussed in his paper, is that the replacement of links by nodes implies that the class of routes admitted includes all those for which, on the original network, no *link* is traversed more than once. The route 6, 4, 2, 1, 3, 4, 5 is admissible because it is a sequence of distinct links (6, 4), (4, 2), (2, 1), (1, 3), (3, 4), (4, 5), even though it is not a chain because node 4 is visited twice.

This gives the justification of the route costs defined in Sect. 9(a), Chap. II, for networks with turn penalties and prohibitions. Following this definition, we can formulate the cheapest route problem as follows. Suppose that $[N; L]$ is a directed network with N the set of nodes 1, 2, 3, ... and L the set of directed links $l_1, l_2, l_3, ...$. Suppose that $c(l_i)$ is the link cost associated with link l_i and $p(l_i, l_j)$ the penalty associated with a turn from link l_i to link l_j. For all l_i, l_j for which there is no permitted turn, $p(l_i, l_j) = \infty$. A finite value of $p(l_i, l_j)$ necessarily implies that $l_i \neq l_j$ and that l_i is directed to and l_j directed from the same node. An *admissible route* on the network can now be formally defined as a sequence $l_1, l_2, ..., l_r$ of links of L which are distinct, except possibly for l_1 and l_r, and are such that $p(l_i, l_{i+1}) < \infty, i = 1, 2, ..., r-1$. The admissible routes include chains (and paths if the network is undirected) as well as the logical routes as illustrated in Fig. 14.4. As defined in (2), Sect. 9, Chap. II, the cost of an admissible route is given by

$$C(l_1, l_r) = \sum_{i=1}^{r-1} [c(l_i) + p(l_i, l_{i+1})].$$ (16)

The problem of determining cheapest admissible routes can be stated in precisely the same form as the cheapest route problem already analyzed for networks without turn penalties and prohibitions. We can take over the theorem and algorithms already described by simply calling nodes links, chains (or paths) admissible routes, and replacing link costs $c(n_i, n_j)$ by $c(l_i) + p(l_i, l_j)$. In particular, we could follow the proof in Appendix A to prove the basic

THEOREM: *The costs of the cheapest admissible routes from link l_1 to link l_r of a network $[N; L]$ with positive link costs $c(l_i)$ and positive turn penalties $p(l_i, l_j)$ are the unique solutions of the functional equations*

$$f(l_1) = 0,$$ (17)

$$f(l_r) = \min_{l_i \neq l_r} [f(l_i) + c(l_i) + p(l_i, l_r)].$$ (18)

Any of the standard cheapest route algorithms can be used to solve these equations.

There is, of course, another obvious way to correctly allow for turn penalties and prohibitions: represent intersections by subnetworks. For example, the METRA computer packages [10] represent an intersection by 8 nodes and as many as 16 links (see Fig. 14.5(a)). If all turning movements are allowed at all intersections, it is more efficient to use a representation of each intersection as 4 nodes and 12 links (see Fig. 14.5(b)). This representation is in fact equivalent to Caldwell's pseudo-network. Although this added network structure removes the

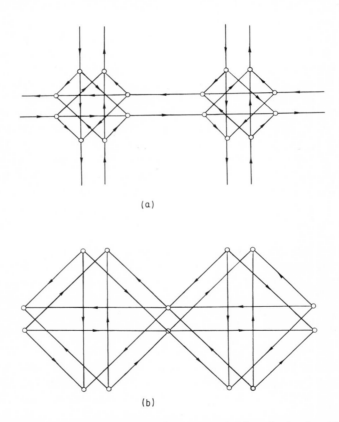

Figure 14.5. Networks with intersections represented by subnetworks. (a) Each intersection represented by 8 nodes and 16 links. (b) Each intersection represented by 4 nodes and 12 links.

problem of turn penalties and prohibitions, it is of limited use in transportation applications because it introduces considerable difficulties
in network coding and greatly increases the demands on computer
storage. For these reasons, the method has not been widely accepted in
computer packages for large networks, except in the METRA programs.

Traffic assignment models make frequent use of cheapest route
algorithms and much effort has been expended in increasing the efficiency of tree-building programs by using the special structure of
transportation networks. For example, it is usual to code main road
networks so that at each node there is a maximum of four outgoing
links, and this restriction greatly facilitates the storing of a network on a
computer. In addition, the network is coded so that no link cost exceeds
a fixed upper bound, and for networks with turn penalties it is usually
sufficient to specify a limited number of turn and intersection types.
Recent tree-building programs, such as the BPR IBM 360 programs
and that described in [11], are extremely efficient.

(d) Cheapest Route Assignment

The most commonly used traffic assignment programs assume that
each traveler chooses the cheapest or perhaps a nearly cheapest route
between his origin and destination. These cheapest routes are determined
by the tree-building algorithms described above, but the efficiency,
power, and accuracy of these algorithms has tended to obscure many
of the difficulties inherent in interpreting cheapest route assignments.
The factors influencing a traveler's choice of routes are likely to be
complex and variable, and research has shown that travel time, distance,
direct and indirect costs, comfort, and convenience all contribute to a
driver's attitude to different routes [12]. To integrate all these factors
into a single link cost, the same for all travelers, is an almost absurd
simplification for a main road network in a metropolitan area. There
may be some agreement about the "cheapest" routes between distant
cities, but drivers certainly have different ideas about the best routes
within a city. In a report [13] of an interesting project carried out in
San Francisco, Jansen gives evidence indicating that between the same
origin and destination drivers followed a variety of routes and the most
popular route was not the shortest nor the quickest! And there is the
obvious difficulty of establishing consistent travel times for the links.
The travel times vary greatly for different times of the day, and even the
fluctuations in successive experimental runs may be large. Added to

this is the difficulty of trying to take measurements for many links of a large network, and it is common practice to rely on rough speed estimates and require the computer to calculate travel times from link distances. It is usually necessary to inspect a sample of the trees to test whether the cheapest routes are logical—if not, the coding of the network is altered. A calculated tree may, on inspection, be found to exclude a route known or predicted to be attractive. To include it in the tree, the speeds on links of the route can be increased or, equivalently, the travel times reduced. This altered coding of the network may improve the particular tree being inspected but other trees may be adversely affected.

The difficulties in cheapest route assignments multiply rapidly as the network size increases. One questions whether the computer packages which can handle thousands of nodes and links can be used by a planner in a meaningful way [14]. From the theoretical and practical points of view, there are considerable advantages in using gross simplified networks for cheapest route assignments.

Of the cheapest route assignment procedures, the *all-or-nothing assignment* is the simplest and most common. Link costs are supposed constant and flow independent, and the traffic between each O–D pair is all assigned to the cheapest route between the pair and none assigned to any other route. For each centroid as home node, the cheapest route trees are loaded from all other centroids back to the home node, accumulating link flows as the calculation proceeds. This loading of the network can be incorporated into the tree-building algorithm. Since the link costs are constant, the all-or-nothing assignment also minimizes the total network cost in accord with the second extremal principle.

The US Bureau of Public Roads Traffic Assignment computer package [8] allows the user the option of choosing a *diversion assignment* instead of the all-or-nothing assignment. Two cheapest route trees are calculated for each centroid, one including freeways in the network and the other excluding them, so that the routes are forced to follow arterial roads. The cheapest freeway route is compared with the cheapest alternate arterial route and a certain proportion of the total interzonal trips are diverted to the freeway route. Various curves and empirical formulas for estimating this diversion have been suggested. The BPR technique uses a time-ratio curve based on the ratio of the time *via* the freeway and the time *via* the quickest alternate route. The diversion varies from 100% to the freeway for a ratio of about 0.5, to 0% for a ratio of about 1.5. For a ratio of 1.0, the diversion is about 42%. The diversion assignment, by allowing more than the one route chosen in all-or-nothing assignment, is more realistic and is well suited for evaluating a new freeway system for a future network. The diversion curves have little theoretical basis and have to be used and interpreted by the planner with considerable care. Some experimental data concerning the diversion of traffic on an arterial road to a parallel toll road has been reported by Michaels [15].

An obvious extension to the all-or-nothing and diversion assignment procedures is realized when allowance is made for multiple routes between origin and destination pairs. For example, the cheapest, next to cheapest, and the next cheapest interzonal routes may be determined and the flow apportioned between these. Burrell [16] has suggested a *multiple route assignment* procedure which gives a single route between an origin and destination but multiple routes between intermediate nodes. A rectangular probability distribution of link costs is assumed, so that each link cost, instead of being constant, can take at random any one of eight equally probable values evenly spread about a mean value. Before each cheapest route tree is built from a home node, the link costs are sampled and the chosen values used in the tree-building algorithm. Because the link costs can differ for each tree, the net effect is to include more links throughout the network for loading with traffic. Although the choice of the distribution of link costs is rather arbitrary, the evidence indicates that the assignments obtained are quite realistic, and the extra computing time over the all-or-nothing assignments is not significant.

15. Minimum Network Cost

As pointed out in the introduction to this chapter, the extremal principles we are considering lead to two important network problems— cheapest routes and minimum network costs. In this section, we shall consider the minimum network cost problem for a network with constant link costs and see how the system-optimized traffic patterns compare with the cheapest route patterns.

We shall begin with a formulation of the problem for a single O–D network using link flows as described in Sect. 8(b), Chap. II, and then reformulate the problem using chain flows as described in Sect. 8(c). Although the conclusions are of course the same, the different approaches throw considerable light on the relation between the two extremal principles we are investigating. In Sect. 15(c) we describe in detail the out-of-kilter algorithm for solving the minimum network cost problem.

(a) Link Flows

In the node–link formulation of the single O–D uncapacitated directed network, we denote the network by $[N; L]$, the nodes by $i = 1, 2, \ldots, n$, origin $i = 1$, destination $i = n$, the directed links by (i, j), the nonnegative link flows by f_{ij}, and constant unit flow costs by $c_{ij} > 0$. As shown in Sect. 8(b), the conservation equations may be written in matrix form as

$$\mathbf{Ef} = \mathbf{g}, \tag{1}$$

where $\mathbf{E}$ is the $n \times l$ node–link incidence matrix, $\mathbf{f}$ the $l \times 1$ link flow vector, and $\mathbf{g}$ the $n \times 1$ O–D flow vector, g being the flow value (assumed a positive integer). In Sect. 9(a), Chap. II, the network cost was defined as

$$C = \sum_{(i,j)} f_{ij} c_{ij}(f_{ij}). \tag{2}$$

For constant link costs and uncapacitated links, the minimum network cost problem can therefore be expressed as the following linear program [17, 18]:

$$\mathbf{f} \geqslant \mathbf{0}, \tag{3}$$

$$\mathbf{Ef} = \mathbf{g}, \tag{4}$$

$$\mathbf{c}^{\mathrm{T}} \mathbf{f} = C(\min), \tag{5}$$

where $\mathbf{c}$ is the $l \times 1$ column with elements c_{ij}. The solution of this LP (linear program) is obvious. If there is a unique cheapest chain from the origin to the destination, with cost $C^*(1, n)$, then $f_{ij} = g$ for each link (i, j) on this chain, $f_{ij} = 0$ for all other links, and the minimum network cost is given by (5) as

$$C^* = gC^*(1, n). \tag{6}$$

If there is more than one cheapest chain, the network flow g can be apportioned between the cheapest chains in any way, and (6) is still true. The essential property of the solution is that the cost of chains which are used for flow are equal and less than or equal to those which are not used.

For our purposes, the significance of this conclusion is that, for the single O–D uncapacitated network with constant link costs, the two extremal principles are equivalent and the user-optimized and system-optimized traffic patterns are equivalent. Also significant is the fact that the cheapest routes naturally arise in the solution to a network flow problem—perhaps a comforting reminder to the reader who missed seeing in the previous section on cheapest routes any reference to flows in networks!

Although the solution of the minimum network cost problem for this special network is intuitive and rather trivial, it is useful to formalize the derivation by an appeal to the duality theory of linear programming (see [17, 18] and Appendix B). This theory associates with the *primal* program (3)–(5), a *dual* program which, in terms of dual variables $-\lambda_i, i = 1, 2, \ldots, n$, can be formulated in matrix form with λ, the $n \times 1$ column with elements λ_i, as

$$-\lambda \quad \text{unrestricted in sign}, \tag{7}$$

$$-\lambda^{\mathsf{T}} \mathbf{E} \leqslant \mathbf{c}^{\mathsf{T}}, \tag{8}$$

$$-\lambda^{\mathsf{T}} \mathbf{g} = V(\max). \tag{9}$$

The significance of the minus sign in the definition of the dual variables will soon appear. By using the explicit form for the matrix $\mathbf{E}$ as given in Sect. 8(b), Chap. II, we can rewrite (7)–(9) as

$$\lambda_i \quad \text{unrestricted in sign}, \qquad i \in N, \tag{10}$$

$$\lambda_j - \lambda_i \leqslant c_{ij}, \qquad\qquad (i, j) \in L, \tag{11}$$

$$g(\lambda_n - \lambda_1) = V(\max). \tag{12}$$

Among important results of duality theory are the following:

(i) if the primal and dual programs have feasible solutions, then they both have optimal solutions and

$$C(\min) = C^* = V(\max) = V^*, \tag{13}$$

and indeed feasible solutions for which $C = V$ are optimal;

(ii) for optimal solutions (distinguished by asterisks), the following *complementary slackness* inferences are valid:

$$\text{if} \quad \lambda_j{}^* - \lambda_i{}^* < c_{ij}, \qquad \text{then} \quad f_{ij}^* = 0, \tag{14}$$

$$\text{if} \quad f_{ij}^* > 0, \qquad \text{then} \quad \lambda_j{}^* - \lambda_i{}^* = c_{ij}; \tag{15}$$

(iii) because there is a linear relation between the constraint (4) of the primal program, one of the dual variables, say $-\lambda_1$, can be taken equal to zero.

In the present context (13), together with (12) and $\lambda_1 = 0$, implies that

$$\lambda_n{}^* = C^*(1,n), \tag{16}$$

i.e., $\lambda_n{}^*$ is the cost of a cheapest O–D chain (and it is for this interpretation that the dual variables were defined with a minus sign).

It also follows that if node i is on a cheapest O–D chain, then $\lambda_i{}^* = C^*(1,i) = $ cost of a cheapest chain from the origin to node i. To see this, it is only necessary to consider

$$\sum (\lambda_j{}^* - \lambda_i{}^*) = \lambda_j{}^* = \sum c_{ij}, \tag{17}$$

where $\sum$ is taken over all links (i,j) in a cheapest O–D chain. This relation forces equality signs in the constraints in (11), i.e.,

$$\lambda_j{}^* - \lambda_i{}^* = c_{ij}. \tag{18}$$

In addition, if $\lambda_j{}^* - \lambda_i{}^* < c_{ij}$, then (i,j) is *not* on a cheapest chain.

When these interpretations are combined with the complementary slackness relations (14) and (15), the solution already described is realized, namely that links not on a cheapest O–D chain get no flow, and an O–D chain with flow is a cheapest chain—and this is merely a restatement of the first extremal principle.

It is not difficult to extend this analysis to the case of a capacitated single O—D network with constant link costs. The primal program is now formulated as

$$\mathbf{f} \geqslant \mathbf{0}, \tag{19}$$

$$\mathbf{f} \leqslant \mathbf{u}, \tag{20}$$

$$\mathbf{E}\mathbf{f} = \mathbf{g}, \tag{21}$$

$$\mathbf{c}^{\mathsf{T}}\mathbf{f} = C(\min), \tag{22}$$

where $\mathbf{u}$ is the $l \times 1$ column with elements u_{ij}, the link capacities. The dual program, with dual variables $-\lambda_i, i \in N$, and $-\mu_{ij}, (i,j) \in L$, becomes

$$\lambda_i \quad \text{unrestricted in sign}, \tag{23}$$

$$\lambda_j - \lambda_i - \mu_{ij} \leqslant c_{ij}, \tag{24}$$

$$\mu_{ij} \geqslant 0, \tag{25}$$

$$g(\lambda_n - \lambda_1) - \sum_{(i,j)} u_{ij}\mu_{ij} = V(\max), \tag{26}$$

for $i \in N$ and $(i,j) \in L$. As before, duality theory implies:

    (i)   the relation

$$\sum c_{ij} f_{ij}^* = g(\lambda_n{}^* - \lambda_1{}^*) - \sum u_{ij} \mu_{ij}^* ; \tag{27}$$

    (ii)   the complementary slackness inferences

$$\text{if}\quad \lambda_j{}^* - \lambda_i{}^* - \mu_{ij}^* < c_{ij}, \qquad \text{then}\quad f_{ij}^* = 0, \tag{28}$$

$$\text{if}\quad \mu_{ij}^* > 0, \qquad\qquad\qquad \text{then}\quad f_{ij}^* = u_{ij}, \tag{29}$$

$$\text{if}\quad f_{ij}^* > 0, \qquad \text{then}\quad \lambda_j{}^* - \lambda_i{}^* - \mu_{ij}^* = c_{ij}, \tag{30}$$

$$\text{if}\quad f_{ij}^* < u_{ij}, \qquad\qquad \text{then}\quad \mu_{ij}^* = 0; \tag{31}$$

   (iii)   that we can take $\lambda_1 = 0$.

The two inferences (30) and (31) combine to yield:

$$\text{if}\quad 0 < f_{ij}^* < u_{ij}, \qquad \text{then}\quad \lambda_j{}^* - \lambda_i{}^* = c_{ij}. \tag{32}$$

Figure 15.1 exhibits the complete relation between $\lambda_j{}^* - \lambda_i{}^*$, μ_{ij}^*, and f_{ij}^*.

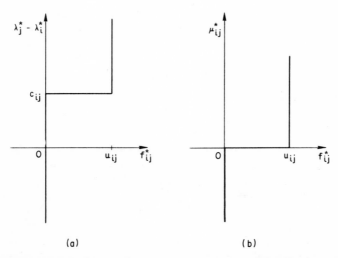

Figure 15.1. Dependence of optimal dual variables on optimal link flows.
(a) $\lambda_i{}^*$ and $\lambda_j{}^*$ are optimal dual variables (node numbers) for nodes i and j, with f_{ij}^* the optimal flow on link (i,j), with link cost c_{ij} and link capacity u_{ij}. (b) μ_{ij}^* is the optimal dual variable (link number) for link (i,j).

To describe the minimum cost or system-optimized traffic pattern and its relation to the user-optimized pattern, we introduce the following terminology. A link (i,j) is called *saturated* if $f_{ij} = u_{ij}$, and *unsaturated* if $f_{ij} < u_{ij}$. For a given network flow, some or all of the flow on a particular O–D chain can be diverted to another chain provided that all the links on the second chain not on the first are unsaturated. Any such chain is said to be *available* for flow from the first chain; otherwise, the chain is said to be *unavailable*. A chain may be available for flow from one chain but unavailable for flow from another. We shall use this concept of available chains to reformulate the first extremal principle.

The route cost for any O–D chain is $\sum c_{ij}$, and by (24),

$$\sum c_{ij} \geqslant \sum (\lambda_j^* - \lambda_i^* - \mu_{ij}^*),$$

that is,

$$\sum c_{ij} \geqslant \lambda_n^* - \sum \mu_{ij}^*, \tag{33}$$

where the summations are over all links of the chain. If a particular chain (1) has positive flow, then by (30)

$$\sum_{(1)} c_{ij} = \lambda_n^* - \sum_{(1)} \mu_{ij}^*. \tag{34}$$

If a second chain (2) is available for flow from this chain, then the links on the second chain not on the first are unsaturated and hence, by (31), $\mu_{ij}^* = 0$ for these links, and

$$\sum_{(1)} \mu_{ij}^* \geqslant \sum_{(2)} \mu_{ij}^*. \tag{35}$$

Combining (33)–(35) therefore gives

$$\sum_{(1)} c_{ij} = \lambda_n^* - \sum_{(1)} \mu_{ij}^* \leqslant \lambda_n^* - \sum_{(2)} \mu_{ij}^* \leqslant \sum_{(2)} c_{ij}. \tag{36}$$

Thus, the route cost for any chain with positive flow is less than or equal to the cost for any chain available for flow from it. This traffic pattern is user-optimized for an extended form of the first principle, modified by inclusion of the concept of available chains. With this modification, we have therefore shown that the system-optimized pattern which minimizes the network cost is a user-optimized pattern.

This extension of the first principle differs from that suggested by Jorgensen [17] who states that "The travel times over routes not used are greater than or equal to the travel times over routes used." In the present context, this would imply that a chain with no flow has a route cost greater than or equal to costs of chains with flow. What must not be overlooked is the possibility that a chain with no flow but smaller route cost may be unavailable because a link is saturated with flow from *other* chain flows.

It is important to note that it has only been shown that the system-optimized traffic pattern is a user-optimized pattern. As will be shown below, the converse is not true—a user-optimized pattern does not necessarily minimize the total network cost.

(b) Chain Flows

We now repeat the above analysis using chain flows instead of link flows. With the notation of Sect. 8(c), Chap. II, the second principle can be stated as the linear program

$$h_j \geqslant 0, \qquad j = 1, ..., m, \tag{37}$$

$$\sum h_j = g, \tag{38}$$

$$f_i = \sum a_{ij} h_j \leqslant u_i, \qquad i = 1, ..., l, \tag{39}$$

$$\sum h_j C_j = C(\min). \tag{40}$$

With dual variables v and $-\mu_i$, the dual program is

$$v \qquad \text{unrestricted in sign}, \tag{41}$$

$$\mu_i \geqslant 0, \qquad i = 1, ..., l, \tag{42}$$

$$v - \sum_{(j)} \mu_i \leqslant C_j, \qquad j = 1, ..., m, \tag{43}$$

$$vg - \sum \mu_i u_i = V(\max), \tag{44}$$

where we adopt the convention that $\sum_{(j)}$ signifies summation over all links on the chain j. For optimal solutions $h_j{}^*$ of the primal (with corresponding optimal link flows $f_i{}^*$) and the optimal solutions $v^*, \mu_i{}^*$ of the dual, duality theory implies

$$C^* = V^*, \tag{45}$$

that is

$$\sum h_j{}^* C_j = v^* g - \sum \mu_i{}^* u_i, \tag{46}$$

as well as the complementary slackness inferences

$$\text{if} \quad h_j{}^* > 0, \qquad\qquad \text{then} \quad v^* - \sum_{(j)} \mu_i{}^* = C_j, \tag{47}$$

$$\text{if} \quad v^* - \sum_{(j)} \mu_i^* < C_j, \qquad\qquad \text{then} \quad h_j^* = 0, \qquad\qquad (48)$$

$$\text{if} \quad f_i^* = \sum a_{ij} h_j^* < u_i, \qquad\qquad \text{then} \quad \mu_i^* = 0, \qquad\qquad (49)$$

$$\text{if} \quad \mu_i^* > 0, \qquad\qquad\qquad \text{then} \quad f_i^* = \sum a_{ij} h_j^* = u_i. \qquad (50)$$

It is easy to show that this system-optimized chain flow pattern is user-optimized in accord with the extended first extremal principle. Suppose, for example, that chain j has positive flow. Then (47) implies

$$v^* - \sum_{(j)} \mu_i^* = C_j. \qquad\qquad (51)$$

If j' is a chain available for flow from j, then the links on j' not in j are unsaturated, and hence, by (49), the corresponding values of μ_i^* are zero, giving

$$\sum_{(j)} \mu_i^* \geqslant \sum_{(j')} \mu_i^*. \qquad\qquad (52)$$

From (43) and (51), it therefore follows that

$$C_j = v^* - \sum_{(j)} \mu_i^* \leqslant v^* - \sum_{(j')} \mu_i^* \leqslant C_{j'}, \qquad (53)$$

that is

$$C_j \leqslant C_{j'}, \qquad\qquad (54)$$

as required for a system-optimized pattern.

This analysis can be illustrated by the network given in Fig. 15.2. The given link capacities and costs are listed in Table 15.1 and the chains are enumerated with their route costs in Table 15.2. For flow value $g = 9$, the chain flow pattern given in Table 15.2 with the consequent link flow

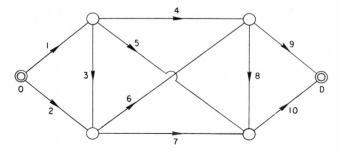

Figure 15.2.   Directed single O–D network. The links are numbered 1, 2, ..., 10.

TABLE 15.1 Link Data for Fig. 15.2

Link number i	Link capacity u_i	Link cost c_i	Link[a] flow f_i^*	$f_i^\dagger$	Dual variable μ_i^*
1	6	1	6	6	3
2	4	5	3	3	0
3	4	1	0	3	0
4	6	3	6	3	0
5	1	7	0	0	0
6	3	1	3	3	1
7	4	6	0	3	0
8	6	3	6	3	0
9	4	6	3	3	0
10	6	3	6	6	0

[a] The system-optimized pattern is not unique. Another optimal solution is $f_8^* = 5$, $f_9^* = 4$, $f_{10}^* = 5$ with all other f_i^* as given in the table.

TABLE 15.2 Chain Data for Fig. 15.2

Chain m_j	Links i	Route cost C_j	Chain flow h_j^*	$h_j^\dagger$	$\sum_{(j)} \mu_i^*$	$v^* - \sum_{(j)} \mu_i^*$
m_1	1, 3, 6, 8, 10	9	0	2	4	9
m_2	1, 3, 6, 9	9	0	1	4	9
m_3	1, 3, 7, 10	11	0	0	3	10
m_4	1, 4, 8, 10	10	4	1	3	10
m_5	1, 4, 9	10	2	2	3	10
m_6	1, 5, 10	11	0	0	3	10
m_7	2, 6, 8, 10	12	2	0	1	12
m_8	2, 6, 9	12	1	0	1	12
m_9	2, 7, 10	14	0	3	0	13

pattern given in Table 15.1 is a system-optimized pattern which minimizes the total network cost C. This is easily verified from the listed values of the dual variables which, with $v^* = 13$, are feasible, satisfying (41)–(43) and giving a value of $V = (13)(9) - 21 = 96$. This is equal to the value of $C = (4)(10) + (2)(10) + (2)(12) + (1)(12) = 96$. These feasible primal and dual solutions with equal values for the objective functions are therefore optimal, as indicated by the use of asterisks.

The chains available for flow from chains with flow are:

Chains m_j with flow	Available chains $m_{j'}$
j	j'
4	3, 5, 6, 9
5	—
7	8, 9
8	—

and it can easily be checked that in all cases $C_j \leqslant C_{j'}$. Note, however, that chains m_1, m_2, m_3, m_6, with costs less than the costs of some chains with flow, have no flow.

It is instructive to check the complementary slackness relations:

(47) is satisfied for $j = 4, 5, 7, 8$;

(48) is satisfied for $j = 3, 6, 9$;

(49) is satisfied for $i = 2, 3, 5, 7, 9$;

(50) is satisfied for $i = 1, 6$.

But the converses of (47) and (48) are contradicted for $j = 1$, and the converses of (49) and (50) are contradicted for $i = 4$.

It is also worth noting that the node–link formulation of this problem requires, in place of the single dual variable v, a dual variable λ_i for each node. The single variable v also has the advantage of an immediate interpretation related to route costs.

Although the system-optimized pattern given by $h_j{}^*$ is a user-optimized pattern, it is simple to illustrate that the converse is not true. In Tables 15.1 and 15.2, we have listed link flows $f_i{}^\dagger$ and chain flows $h_j{}^\dagger$ which are also user-optimized. This can be verified from the chains available for flow from chains with flow:

Chains m_j with flow	Available chains $m_{j'}$
j	j'
1	2, 3, 4, 5, 6, 7, 8, 9
2	5
4	3, 5, 6, 9
5	—

It can easily be checked that in all cases $C_j \leqslant C_{j'}$. But the network cost is

$$C^\dagger = (2)(9) + (1)(9) + (1)(10) + (2)(10) + (3)(14) = 99 > C^* = 96 \,.$$

This example also shows that different user-optimized patterns may give different network costs.

It is perhaps surprising that the optimal solution $h_j{}^*$ does not use either of the cheapest routes m_1, m_2. If travelers were to choose the cheapest routes on a first come, first served basis, a solution the same as $h_j{}^\dagger$ would be obtained (with possible unimportant swaps between routes of equal cost).

We summarize this analysis by enunciating the *principle of available chains:*

For a capacitated network with constant link costs a system-optimized traffic pattern (minimizing the total network cost) is a user-optimized pattern in the sense that the chain costs for any chain with positive flow is less than or equal to the cost for any other chain available for flow from it. Different user-optimized patterns may have different network costs, so that a user-optimized pattern is not necessarily a system-optimized pattern.

(c) The Out-of-Kilter Algorithm

In this part we present a finite algorithm which has enjoyed much success in the realm of practical computation, and also gives substantial insight into the nature of the physical flow process itself. For reasons we will expand upon, it has come to be known as the *Out-of-Kilter* algorithm (see [1], Sect. 11, pp. 162–169).

The algorithm that we describe is also special, in the sense that we restrict ourselves to minimum cost flow problems where one can begin computations with a set of feasible flows, i.e., nonnegative link flows that satisfy the conservation laws and are less than the flow capacity explicitly stated for each link. The algorithm will be described in terms of the link flow formulation of the minimum network cost problem given in Sect. 15(a), and it consists of three distinct parts:

(i)   the identification of the state of a link and its appropriate kilter number,
(ii) a subroutine for rerouting flow,
(iii) a subroutine for making dual variable changes.

As we will prove, this algorithm is finite and terminates with an optimal solution of the minimum cost flow problem stated in (19)–(22). We also assume that all c_{ij}, u_{ij}, and g_i data are integers.

(i)   STATES AND KILTER NUMBERS

We associate with each link in the network a complementary slackness diagram, Fig. 15.3, similar to Fig. 15.1(a), in which the vertical axis is

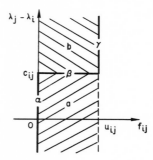

Figure 15.3.   The complementary slackness diagram with associated states.

State	Link flow	Dual variables	
α	$f_{ij} = 0$	$\lambda_j - \lambda_i < c_{ij}$	⎫
β	$0 \leqslant f_{ij} \leqslant u_{ij}$	$\lambda_j - \lambda_i = c_{ij}$	⎬ In-kilter
γ	$f_{ij} = u_{ij}$	$\lambda_j - \lambda_i > c_{ij}$	⎭
a	$0 < f_{ij} \leqslant u_{ij}$	$\lambda_j - \lambda_i < c_{ij}$	⎫ Out-of-kilter
b	$0 \leqslant f_{ij} < u_{ij}$	$\lambda_j - \lambda_i > c_{ij}$	⎭

the difference in dual or node variables, and the horizontal axis refers to the flow variable. At each stage of the computation every link in the network can be classified in one of five states, Greek symbols corresponding to the "in-kilter" states which lie on the complementary slackness diagram of Fig. 15.3. Roman letters a and b correspond to "out-of-kilter" states for each link. By "out-of-kilter" we simply mean that the dual variables and flow values for a given link lie to one side of, above, or below but *not on* the heavy lines of Fig. 15.3. Whenever we inspect a link in the network, we can measure the flow through that link and the difference in the dual variables for the corresponding A and B nodes. These lead to a point and thus a state on the complementary slackness diagram.

For links in state a or b we compute positive *kilter numbers*

$$\kappa_a = (f_{ij})(c_{ij} - \lambda_j + \lambda_i), \qquad (55)$$

$$\kappa_b = (u_{ij} - f_{ij})(\lambda_j - \lambda_i - c_{ij}). \qquad (56)$$

In other words, κ_a and κ_b are products of distances from horizontal and vertical segments of the complementary slackness diagram in Fig. 15.3. To put it another way, the numbers κ_a and κ_b measure the degree of nonoptimality of out-of-kilter links. In-kilter links in states α, β, or γ have a kilter number $\kappa_\alpha = \kappa_\beta = \kappa_\gamma = 0$ and are optimal.

(ii) FLOW-REROUTING SUBROUTINE

In this subroutine we attempt to locate meshes that include at least one out-of-kilter link whose kilter number can be decreased by increasing or decreasing the flow in such a way as to preserve feasibility of all flows in the network, and simultaneously ensure that no kilter numbers of the links in the mesh are increased. To decrease kilter numbers of out-of-kilter links, we must reduce flows on links in state a and increase flows on links in state b. To maintain zero kilter numbers for in-kilter links, we must ensure that links in states α, β or γ do not move into state a or b. Dual variables on nodes are held fixed in this phase of the computation.

To locate such a mesh, let us first concentrate on an out-of-kilter link (s, t) in state b. In this case, we can obviously reduce its kilter number, given by (56), by increasing the flow in the link. To increase the flow

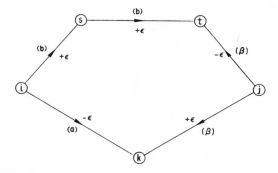

Fig. 15.4. Mesh with out-of-kilter link and flow-augmenting path. (s, t) is the out-of-kilter link in state b; (j, k), (i, s) are forward links; (j, t), (i, k) are two reverse links in the flow-augmenting path from t to s. Initial states of the links and the additional flow assigned are indicated.

on (s, t) and maintain feasible flows in the rest of the network, we search for any *return path* with origin node t and destination node s (see Fig. 15.4), such that forward links in that path can have their flows increased and reverse links in that path can have their flows decreased while observing the principle we mentioned in the previous paragraph: either kilter numbers on out-of-kilter links decrease or kilter numbers on in-kilter links do not increase. If such a return path can be found, we call it a *flow-augmenting path*. A link in the typical return path from t to s may be in one of five states: α, β, γ, a, or b, and it may be a forward or reverse link. It appears there are ten distinct possibilities that we have to consider, but fortunately we only need to consider four. The four types of links which can be, but are not necessarily, members of a flow-augmenting path are given in Table 15.3.

TABLE 15.3   LINK FLOWS AND STATES ON RETURN PATH

Current state	Current flow	Link type	Attempt to:	New state
β	$< u_{ij}$	forward	increase flow	β
β	> 0	reverse	decrease flow	β
a	> 0	reverse	decrease flow	a or α
b	$< u_{ij}$	forward	increase flow	b or γ

Notice that states α and γ are excluded because *any* flow changes on such links would either increase the kilter numbers of those links or lead to infeasible flows. Similar remarks apply to forward links in state a (an increase of flow would increase the kilter number) and reverse links in state b (a decrease in flow would increase the kilter number). The possible change in flow for links in Table 15.3 is also shown in Fig. 15.5(a) and (b).

If a flow augmenting path is found, at least one unit of flow can be added to the out-of-kilter link (s, t) which initiated the computation and all the other links in the return flow-augmenting path from t to s. The net effect of this flow change is to increase by at least one unit the quantity of flow moving around the mesh. In general, more than one unit of flow can be added to the out-of-kilter link with a maximum value obtained by the following computation:

On forward links of the flow-augmenting path, flows must not exceed the capacity of a single link; thus, the increase in flow is a number less than or equal to

$$\varepsilon_1 = \min_{\substack{\text{forward} \\ \text{links}}} \{u_{ij} - f_{ij}\}. \tag{57}$$

On reverse links of the flow-augmenting path the flow reduction must not lead to negative flows; thus, it is a number less than or equal to

$$\varepsilon_2 = \min_{\substack{\text{reverse} \\ \text{links}}} \{f_{ij}\}. \tag{58}$$

Finally, one must not exceed the capacity of the out-of-kilter link that initiated the computation. Thus, the maximum flow increase that is feasible throughout the mesh is

$$\varepsilon = \min \{u_{st} - f_{st}; \varepsilon_1; \varepsilon_2\}. \tag{59}$$

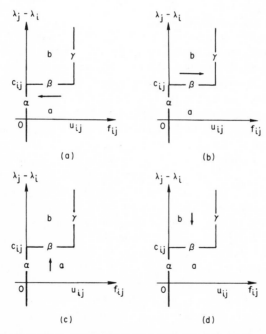

Figure 15.5. The effect of flow and dual variable changes. (a) Decreasing flow on links in state a or β. (b) Increasing flow on links in state b or β. (c) Increasing the difference in node numbers on links in state a or α. (d) Decreasing the difference in node numbers on links in state b or γ.

Increasing the flow of all forward links (including the out-of-kilter link (s, t)) and reducing the flow of all reverse links by ε is, first of all, feasible for all links in the mesh and secondly, reduces the kilter number of at least one link in the mesh. Flows and kilter numbers of links not on that mesh will be unchanged.

If, initially, the out-of-kilter link (s, t) is in state a, we use node s as an origin and attempt to find a flow-augmenting path leading to the destination node t. Again we restrict links on flow-augmenting paths to those listed in Table 15.3. If such a path can be found, it is, therefore, possible to reduce the flow, and hence the kilter number, on at least link (s, t). In this case, the maximum flow reduction that can be allowed is again given by (57) and (58), and

$$\varepsilon = \min \{f_{st}; \varepsilon_1; \varepsilon_2\}. \qquad (60)$$

To illustrate a flow-augmenting path for this case, one can alter Fig. 15.4 by interchanging states a and b for the links (s,t), (i,s), (i,k) and by reversing the signs in front of ε.

Let us consider, in a little more detail, what happens in those cases where no flow-augmenting path can be found leading from the origin node back to the destination node. As one follows a sequence of nodes and distinct links from the origin node, one must reach an intermediate node from which no link of the type listed in Table 15.3 either enters or leaves. Conceptually, one can think of that intermediate node as the last node on a path from the origin, all of whose nodes are labeled; nodes not reachable from the labeled nodes by means of links of the type listed in Table 15.3 are unlabeled. To put it another way, nodes connected to labeled nodes by paths that consist only of links listed in Table 15.3 can also be labeled. Nodes that cannot be connected to labeled nodes by paths that include only links listed in Table 15.3 are unlabeled. It can be proved (although we do not do so here) that if no flow-augmenting path exists, the labeled nodes include the origin node, the unlabeled nodes include the destination node, and the links leading from labeled to unlabeled nodes form a cut-set (see Sect. 9, Chap. II). Once an out-of-kilter link (s, t) has been selected (either state a or b), the process of searching for paths and reassigning flows terminates in one of two ways: either kilter numbers of all links are identically zero, in which case we have reached an optimum and the algorithm terminates, or no such flow-augmenting path can be found; in the latter case, the algorithm for finding flow-augmenting paths automatically constructs a cut-set $(X, \overline{X})$ (see Sect. 9, Chap. II) of labeled and unlabeled nodes, such that the origin node is in X, the destination node is in $\overline{X}$, and forward links leading from nodes in X to nodes in $\overline{X}$ or reverse links from $\overline{X}$ to X can be grouped into exclusive subsets:

(i) forward links in $(X, \overline{X})$ are in one of two subsets: either they are in state β or γ with $\lambda_j - \lambda_i \geqslant c_{ij}$ and $f_{ij} = u_{ij}$, or in a set L_1 containing links in state a or α, with $\lambda_j - \lambda_i < c_{ij}$. In other words, forward links in state a or α define the set L_1;

(ii) reverse links in $(\overline{X}, X)$ are in one of two subsets: either they are in state β or α with $\lambda_j - \lambda_i \leqslant c_{ij}$ and $f_{ij} = 0$, or in a set L_2 containing links in state b or γ, with $\lambda_j - \lambda_i > c_{ij}$. Reverse links in state b or γ define the set L_2.

Once we have observed the fact that no flow-augmenting path can be found leading from the origin node back to the destination node of an out-of-kilter link and have identified the members of L_1 and L_2, we then turn our attention to a subroutine which changes the value of the dual variables on the nodes in X and $\overline{X}$. As we will see, dual variable changes which reduce kilter numbers of links in L_1 and L_2 may also help us find additional flow-augmenting paths.

Let us review the procedure that has been described. We locate an out-of-kilter link and attempt to find a flow-augmenting path with either the A or the B node of that link being the origin, the other node being the destination node of the path. A flow-augmenting path has the property that at least one unit of flow can be added to forward links, removed from reverse links of the path, and, depending upon the type of out-of-kilter link selected initially, either added or subtracted from the latter in order to preserve feasibility of flows throughout the entire network. The flow-augmenting path and hence the mesh that includes the initial out-of-kilter link has the further property that the kilter number of one or more links on the mesh can be reduced. Note that a large number of paths, and hence meshes, that include the out-of-kilter link, will not, in general, provide flow-augmenting paths. An example of such a path might be one which included only α and γ links. It is obvious, from the list contained in Table 15.3, that such links are inadmissible candidates for a flow-augmenting path because they violate the principle of selecting links that do not increase kilter numbers. If no flow-augmenting path can be found, then the nodes of the transportation network can be partitioned into two disjoint subsets, X and $\overline{X}$, with the origin in X and the destination in $\overline{X}$. All forward links from X to $\overline{X}$ define a cut-set;

a subset of these and of the reverse links leading from $\overline{X}$ to X are used to calculate dual variable changes on the nodes in order to provide either additional flow-augmenting paths or termination of the algorithm at its optimum value. These dual variable changes are the subject of the next part.

(iii)   DUAL VARIABLE CHANGE SUBROUTINE

As we have mentioned, the nodes of the transportation network can now be partitioned into two disjoint sets, X and $\overline{X}$. If one can think of the previous subroutine as providing a series of computations whose objective is to move nonoptimal link states in the horizontal directions indicated in Fig. 15.5(a) and (b), then the objective of the subroutine now to be described can be stated as that of moving nonoptimal states in the vertical directions indicated by Fig. 15.5(c) and (d). Just as we did not allow flow changes to be made on α or γ links in the previous section, we will not allow β links with nonzero flows less than capacity to enter into the computation for new node numbers. To understand why this must be so, consider the following new node or dual variable numbers (primes denote new):

$$
\begin{aligned}
\lambda_i' &= \lambda_i, & i \in X, \\
&= \lambda_i + \delta, & i \in \overline{X}, \quad \delta > 0.
\end{aligned}
\tag{61}
$$

Although we do not yet know what value to use for δ, we see that the difference in new dual variables for each link in the network can be written in terms of the old dual variables as follows:

$$
\begin{aligned}
\lambda_j' - \lambda_i' &= \lambda_j - \lambda_i, & \{(i,j) \in L \,|\, i \in X, j \in X\} \\
&= \lambda_j - \lambda_i, & \{(i,j) \in L \,|\, i \in \overline{X}, j \in \overline{X}\} \\
&= \lambda_j - \lambda_i + \delta, & \{(i,j) \in L \,|\, i \in X, j \in \overline{X}\} \\
&= \lambda_j - \lambda_i - \delta, & \{(i,j) \in L \,|\, i \in \overline{X}, j \in X\}.
\end{aligned}
\tag{62}
$$

Furthermore, as long as δ is positive then the states of *all* links leading from nodes X to nodes in $\overline{X}$ always move in the direction indicated by Fig. 15.5(c) and the states of *all* links leading from nodes in $\overline{X}$ to nodes in X move in the direction indicated by Fig. 15.5(d). In both cases the direction is toward, not away from, the complementary slackness diagram. A β link with nonzero flow less than capacity would move out-of-kilter and, therefore, must be excluded.

Since we do not want a or α links in L_1 to cross the β state and move into the nonoptimal state b, nor do we want the b or γ links in L_2 to move into the nonoptimal state a, we can compute

$$\delta_1 = \min_{(i,j)\in L_1} \{c_{ij} - \lambda_j + \lambda_i\}, \tag{63}$$

$$\delta_2 = \min_{(i,j)\in L_2} \{\lambda_j - \lambda_i - c_{ij}\}, \tag{64}$$

and choose δ so that

$$\delta = \min \{\delta_1 ; \delta_2\}. \tag{65}$$

Geometrically, δ is the smallest absolute vertical distance of the state of any forward link in state a or α or any reverse link in b or γ from the horizontal β state in the complementary slackness diagram of each link. Thus, the computations in (63)–(65) ensure that we do not cross the optimal state, and not only guarantee that the out-of-kilter link (s, t) will have a reduced kilter number, but also that in-kilter links in states α, β, or γ will remain in-kilter. Of course this change of node numbers may also reduce the kilter number of many links besides that of (s, t).

Once the new dual variables have been calculated, we return to the flow rerouting subroutine of (ii) and look for an out-of-kilter link and flow-augmenting paths as before.

(iv) DECREASE IN KILTER NUMBERS AND FINITENESS OF ALGORITHM

In this section we show that any step which includes the flow-rerouting and dual variable change subroutines decreases at least one kilter number or leads to an optimal solution.

In the flow-rerouting subroutine we focus attention on the out-of-kilter link (s, t) that is initially selected. If this link is in state b then, by definition (56),

$$\lambda_t - \lambda_s > c_{st}; \qquad f_{st} < u_{st}; \qquad \kappa_b = (u_{st} - f_{st})(\lambda_t - \lambda_s - c_{st}) > 0. \tag{66}$$

If a flow-augmenting path is found with a consequent $\varepsilon > 0$ increase in mesh flow, the kilter number of link (s, t) changes by the amount

$$\kappa_b' - \kappa_b = (-f_{st}' + f_{st})(\lambda_t - \lambda_s - c_{st}) = -\varepsilon(\lambda_t - \lambda_s - c_{st}) < 0. \tag{67}$$

On the other hand, if link (s, t) is initially in state a then, by definition (55),

$$\lambda_t - \lambda_s < c_{st}; \qquad f_{st} > 0; \qquad \kappa_a = f_{st}(c_{st} - \lambda_t + \lambda_s) > 0, \tag{68}$$

and a flow-augmenting path with an $\varepsilon > 0$ decrease in mesh flow changes the kilter number by the amount

$$\kappa_a' - \kappa_a = (f_{st}' - f_{st})(c_{st} - \lambda_t + \lambda_s) = -\varepsilon(c_{st} - \lambda_t + \lambda_s) < 0. \qquad (69)$$

Of course the kilter numbers of other nonoptimal links on the mesh will also decrease in proportion to ε.

If no flow-augmenting path can be found, and kilter numbers of all links in the network are not all zero, we have either $t \in X$, $s \in \bar{X}$ $((s,t)$ in state b), or $s \in X$, $t \in \bar{X}$ $((s,t)$ in state a). In the former case, the change in kilter number of link (s, t) due to the dual variable change given by (61) is

$$\kappa_b' - \kappa_b = (u_{st} - f_{st})[(\lambda_t' - \lambda_s' - c_{st}) - (\lambda_t - \lambda_s - c_{st})]$$

$$= (u_{st} - f_{st})[(\lambda_t' - \lambda_t) - (\lambda_s' - \lambda_s)]$$

$$= -\delta(u_{st} - f_{st}) < 0, \qquad (70)$$

while in the latter case, the dual variable change yields

$$\kappa_a' - \kappa_a = f_{st}[(\lambda_s' - \lambda_s) - (\lambda_t' - \lambda_t)] = -\delta f_{st} < 0. \qquad (71)$$

In other words, the kilter number is a strictly decreasing function in any step which includes both the flow rerouting subroutine and the dual variable change. All data for c_{ij}, u_{ij}, and g_i is integral; hence, all flow and dual variable changes, ε and δ, are also integral and the algorithm must terminate in a finite number of steps with all kilter numbers identically zero. Thus, the complementary slackness conditions are satisfied, and the algorithm obtains necessary and sufficient conditions for an optimum solution of the minimum cost flow problem in (19)–(22).

(v)   EXAMPLE

To illustrate the application of the out-of-kilter algorithm, we shall consider the network shown in Fig. 15.2, with link capacities and costs as listed in Table 15.1. For a flow value $g = 9$, we take as initial feasible flows the link flows $f_i^\dagger$ listed in Table 15.1, and for node numbers we choose those shown in Fig. 15.6. The initial link states are listed in column 4 of Table 15.4.

For the flow rerouting subroutine, we choose the out-of-kilter link 8 in state b, and note that the flow in mesh 8, 7, 3, 4 can be increased by an amount $\varepsilon = 3$. The new link flows in the network are listed in column 5

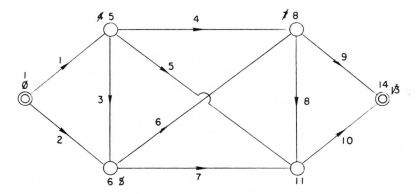

Figure 15.6. Network for out-of-kilter example. The network is the same as in Fig. 15.2 with the link capacities and costs as in Table 15.1. Links are numbered 1, 2, ..., 10. The numbers on the nodes are the dual variables used in the out-of-kilter algorithm. The initial node numbers are 0, 4, 5, 7, 11, 13, and the final numbers 1, 5, 6, 8, 11, 14.

TABLE 15.4 OUT-OF-KILTER EXAMPLE

Link number	Initial link flow	Initial diff. in node nos.	Initial link state	New link flow	New link state	New diff. in node nos.	Final link state
1	6	4	γ	6	γ	4	γ
2	3	5	β	3	β	5	β
3	3	1	β	0	β	1	β
4	3	3	β	6	β	3	β
5	0	7	β	0	β	6	α
6	3	2	γ	3	γ	2	γ
7	3	6	β	0	β	5	α
8	3	4	b	6	γ	3	β
9	3	6	β	3	β	6	β
10	6	2	a	6	a	3	β

of Table 15.4 and the new link states in column 6. Link 10 is still out-of-kilter, but no further flow-augmenting paths are available, and the subroutine ends with

$$(X, \overline{X}) = \{10\}, \qquad L_1 = \{10\}, \tag{72}$$

$$(\overline{X}, X) = \{5, 7, 8\}, \qquad L_2 = \{8\}. \tag{73}$$

We now proceed to the dual variable change subroutine. An increase by $\delta = 1$ to the node numbers in $\overline{X}$ gives the final node numbers in Fig. 15.6; in fact, this completes the calculation. The new differences in node numbers are listed in column 7 of Table 15.4 and the new link states in the last column. All links are in-kilter so that the optimal flow has been obtained.

Even this simple example suggests that hand calculations using the out-of-kilter algorithm are extremely tedious for large networks. Fortunately, the algorithm is well suited for computer calculations and very efficient programs are readily available [19].

16.   Flow Dependent Costs

We have shown in the previous two sections that when costs are flow independent, the two extremal principles lead to user-optimized and system-optimized traffic patterns which are equivalent. If there are no capacity constraints, the network flows are obtained by an "all-or-nothing" assignment in which all flow between each O–D pair is assigned to the cheapest chain connecting the pair. We found that when capacity constraints are explicitly included so that links can become saturated, the first principle had to be modified to include the concept of chains available for flow. With this modification we found that, as summarized in the principle of available chains, a system-optimized pattern is a user-optimized pattern but in general there are user-optimized patterns which are not system-optimized ones.

In studying more realistic traffic assignment problems, it is imperative to take into account the effects of traffic congestion by allowing travel costs to increase with traffic flow. The introduction of flow dependent costs considerably complicates the analysis of the extremal principles, but we shall derive some interesting results for the resulting flow patterns. We shall carefully define user-optimized and system-optimized traffic patterns and state and prove two theorems giving necessary and sufficient conditions for such patterns. We shall also show that at least one user-optimized pattern for a given network is the system-optimized pattern for an associated network problem with related link costs.

In our analysis we shall consider the general case of multiple O–D networks. We shall assume flows to be continuous rather than discrete integer variables and we shall exclude the possibility that a link can become saturated. We may allow for a capacitated link by restricting the link flow to be less than a link capacity but we specifically exclude the possibility of equality. This assumption saves us from being concerned with the concept of available chains. We shall also impose restrictions on the flow-dependent cost functions, but these will be stated later.

(a) Multicommodity Formulation

Suppose we have an enumeration of chains from each origin centroid $O^{(k)}$ to each destination centroid $D^{(k)}$, and a set of traffic flows $g^{(k)}$, one from each origin $O^{(k)}$ to each destination $D^{(k)}$, of a multi O–D transportation network. In principle, the traffic assignment problem is to assign the $g^{(k)}$ to the chains according to some cost criteria. The link flows f_i can then be found by aggregating the appropriate chain flows $h_j^{(k)}$ and indeed we restrict ourselves to link flow patterns which can be obtained this way. As in (26) and (30), Sect. 8, Chap. II, the relevant conservation equations are

$$g^{(k)} = \sum_j h_j^{(k)}; \qquad h_j^{(k)} \geqslant 0, \tag{1}$$

$$f_i = \sum_k \sum_j a_{ij}^{(k)} h_j^{(k)}, \tag{2}$$

where $a_{ij}^{(k)}$ are elements of the link-chain matrix for the origin–destination pair $O^{(k)}$–$D^{(k)}$. For each $O^{(k)}$–$D^{(k)}$ pair and each chain connecting this pair of nodes, we associate a corresponding kth commodity average chain cost. We denote the set of chains for the kth commodity by $M^{(k)}$, so that

$$C_j^{(k)} = C_j^{(k)}(\mathbf{h}) = \sum_i a_{ij}^{(k)} c_i(f_i), \qquad j \in M^{(k)}. \tag{3}$$

As before, $\mathbf{h}$ denotes the vector of all chain flows or the chain flow traffic pattern and $c_i(f_i)$ is the average cost to an individual using the ith link when the total link flow, as in (2), is the sum of all commodity flows through the ith link. It is important to stress the fact that in general $C_j^{(k)}$ depends on $\mathbf{h}$ and not just the chain flows associated with commodity k. The traffic assignment problems that we study involve finding user-optimized flow patterns satisfying certain inequalities among the route costs of (3) and system-optimized patterns minimizing the total cost function

$$C = C(\mathbf{h}) = \sum_i f_i c_i(f_i) \tag{4}$$

subject to the conservation equations (1) and (2).

(b) Equilibrium Flow Patterns for Noncooperative Users

It is important to obtain a more precise understanding of the user-optimized traffic patterns which are the equilibrium patterns achieved by individual users of the transportation network who try to minimize their own route costs without regard to other users.

Consider two chains r and s leading from $O^{(k)}$ and $D^{(k)}$ and two distinct assignments of chain flow, the first corresponding to a chain flow pattern **h** with $h_r^{(k)} > 0, h_s^{(k)} \geqslant 0$ the second corresponding to a flow pattern **h′** with

$$h_r'^{(k)} = h_r^{(k)} - \Delta \geqslant 0,$$
$$\qquad\qquad\qquad\qquad r, s \in M^{(k)}, \qquad (5)$$
$$h_s'^{(k)} = h_s^{(k)} + \Delta > 0,$$

with all other $h_j'^{(k)} = h_j^{(k)}$ and $0 < \Delta \leqslant h_r^{(k)}$. The corresponding link flow patterns are **f** and **f′**. In other words a flow of Δ has been reassigned from a chain r to another chain s. The reassignment is feasible because $\mathbf{h}' \geqslant 0$ and (1) is satisfied. The chain costs for the original flow pattern are obtained directly from (3):

$$C_r^{(k)} = C_r^{(k)}(\mathbf{h}) = \sum_i c_i(f_i) a_{ir}^{(k)}, \qquad (6)$$

$$C_s^{(k)} = C_s^{(k)}(\mathbf{h}) = \sum_i c_i(f_i) a_{is}^{(k)}. \qquad (7)$$

Similarly, for the primed case, we have

$$C_r'^{(k)} = C_r^{(k)}(\mathbf{h}') = \sum_i c_i(f_i') a_{ir}^{(k)}, \qquad (8)$$

$$C_s'^{(k)} = C_s^{(k)}(\mathbf{h}') = \sum_i c_i(f_i') a_{is}^{(k)}. \qquad (9)$$

In (8) and (9) we must interpret the primed link flows as follows (see Fig. 16.1):

$$f_i' = f_i - \Delta \qquad \text{if link } i \text{ is in chain } r, \text{ not in chain } s,$$
$$\text{i.e., } a_{ir}^{(k)} = 1, \quad a_{is}^{(k)} = 0$$
$$= f_i \qquad \text{if link } i \text{ is in chain } r \text{ and } s,$$
$$\text{i.e., } a_{ir}^{(k)} = 1, \quad a_{is}^{(k)} = 1$$
$$= f_i + \Delta \qquad \text{if link } i \text{ is in chain } s, \text{ not in chain } r,$$
$$\text{i.e., } a_{ir}^{(k)} = 0, \quad a_{is}^{(k)} = 1. \qquad (10)$$

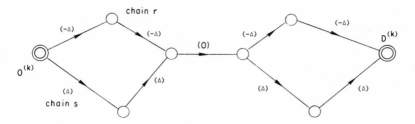

Figure 16.1. The reassignment of flow from one chain to another. A flow Δ on chain r from origin $O^{(k)}$ to destination $D^{(k)}$ is reassigned to chain s between the same O–D pair. The consequent changes in link flows are indicated in parentheses.

In other words the effect of the reassigned flow Δ on chains r and s is to cancel one another on links common to both chains. We say that a *user-optimized traffic pattern* has been reached when for every pair of chains r and s (k fixed), for *every* feasible $\Delta > 0$, and for every commodity k, the following inequalities hold

$$C_r^{(k)} = C_r^{(k)}(\mathbf{h}) \leqslant C_s^{(k)}(\mathbf{h}') = C_s'^{(k)} \qquad r,s \in M^{(k)}. \tag{11}$$

The meaning of the inequalities in (11) is simply this. Consider a number of travellers using a particular route (chain r). In particular this means that the chain flow is strictly positive. A user-optimized traffic pattern $\mathbf{h}$ has been reached if for feasible reassignments of any travellers from r to *any other* chain s (connecting the same O–D pair) the average cost of every s under the new traffic pattern is greater than or equal to the average cost of r under the old. While the costs of chains r and s are being compared, the chain flows due to travellers using routes other than those in the set $M^{(k)}$ are held fixed. And remember that since flow variables are assumed continuous, travellers are infinitely divisible! Thus, the user-optimized equilibrium we have defined must satisfy local rather than global conditions in the sense that comparisons are restricted to chains between a given O–D pair. It may be useful to point out that there are potentially an infinite number of such inequalities (because Δ can be any feasible reassignment) and that not only are we comparing different chains r and s in (11) but we are also comparing different flow patterns: $\mathbf{h}$ on the left and $\mathbf{h}'$ on the right-hand side of the inequality.

In the remainder of this chapter we will assume that $f_i c_i(f_i)$ is strictly convex and increasing. As we show in (8), Appendix C, this implies that the average costs, $c_i(f_i)$, are positive and strictly increasing. We are now in a position to derive a necessary and sufficient condition for a flow pattern to be a user-optimized pattern as defined by (11).

THEOREM: *A flow pattern* **h** *is a user-optimized pattern if and only if for every commodity* k *there exists an ordering* $1, 2, ..., p, p+1, ..., m(k)$ *of the chains from* $O^{(k)}$ *to* $D^{(k)}$ *such that:*

$$C_1^{(k)}(\mathbf{h}) = C_2^{(k)}(\mathbf{h}) = \cdots C_p^{(k)}(\mathbf{h}) \leqslant C_{p+1}^{(k)}(h) \cdots \leqslant C_{m(k)}^{(k)}(\mathbf{h})$$

$$h_1^{(k)} > 0; \quad h_2^{(k)} > 0; \quad \cdots h_p^{(k)} > 0; \quad h_{p+1}^{(k)} = 0 \cdots h_{m(k)}^{(k)} = 0. \tag{12}$$

More simply, this condition implies that the chains between any O–D pair can be grouped into two subsets, one for chains with flow, the other for chains without flow. The costs of chains in the first subset are equal and less than or equal to the costs of chains in the second subset.

To show that this condition is *sufficient* we show that any feasible reassignment of chain flows from a chain r, $1 \leqslant r \leqslant p \leqslant m(k)$, to any chain s, $1 \leqslant s \leqslant m(k)$, satisfies the inequalities of (11). With this choice of chains we have from (12) that

$$C_r^{(k)}(\mathbf{h}) \leqslant C_s^{(k)}(\mathbf{h}), \qquad 1 \leqslant r \leqslant p. \tag{13}$$

The assumption that total link costs are strictly convex yields the strict inequality ((8), Appendix C)

$$c_i(f_i') > c_i(f_i) \qquad \text{for every } f_i' > f_i. \tag{14}$$

Since the feasible reassignment of (5) ensures that at least one link in chain s has a strict increase in flow, it follows that

$$C_s^{(k)}(\mathbf{h}') = \sum_i c_i(f_i') a_{is}^{(k)} > C_s^{(k)}(\mathbf{h}) = \sum_i c_i(f_i) a_{is}^{(k)}. \tag{15}$$

Substituting (15) into (13) gives the strict inequality

$$C_r^{(k)}(\mathbf{h}) < C_s^{(k)}(\mathbf{h}') \tag{16}$$

which satisfies the definition (11) of a user-optimized flow pattern.

To show that the condition is *necessary* we assume that the condition is not satisfied so that there exist two chains r and s with chain costs

$$C_r^{(k)}(\mathbf{h}) > C_s^{(k)}(\mathbf{h}) \qquad \text{and} \qquad h_r^{(k)} > 0. \tag{17}$$

We now deduce that the traffic pattern is *not* user-optimized. The assumption that total link costs are strictly convex implies that the average chain costs are continuous functions of the flow. (See Appendix C.) Hence, we can always pick a sufficiently small feasible reassignment of flow from chain r to chain s (in the sense of (5)) to obtain the inequality

$$C_r^{(k)}(\mathbf{h}) > C_s^{(k)}(\mathbf{h}') > C_s^{(k)}(\mathbf{h}). \tag{18}$$

But this last inequality contradicts the definition (11) of a user-optimized flow pattern. This completes the proof of the theorem.

(c) Minimum Network Cost Flow Patterns

In this section we make use of two additional costs: the marginal cost of flow in a link and the marginal cost of flow in a chain. By marginal cost we mean the change in cost due to a small change in flow, *or* more precisely, the derivative of total cost with respect to flow. The notation we use for marginal costs in the ith link and jth chain is

$$d_i(f_i) = \frac{d}{df_i}[f_i c_i(f_i)] \tag{19}$$

and

$$D_j^{(k)}(\mathbf{h}) = \sum_i d_i(f_i) a_{ij}^{(k)}. \tag{20}$$

Again, the interpretation of the sum in (20) is over all links in the jth chain connecting $O^{(k)}$ to $D^{(k)}$. Throughout this section we assume as before that $f_i c_i(f_i)$ is strictly convex and increasing so that the marginal costs defined by (19) and (20) are positive, continuous, and increasing. Since the flows on links in chain j generally depend on other chain flows, the marginal chain cost in (20) is again written explicitly as a function of the flow pattern, $\mathbf{h}$, on the entire network. The total network cost,

$$C = C(\mathbf{h}) = \sum_i f_i c_i(f_i), \tag{21}$$

sums the total cost of flow in each link over all links in the network. In this section we are interested in deriving a necessary and sufficient condition for a system-optimized flow pattern which minimizes C in (21) subject to the conservation and nonnegativity of chain flows in (1) and (2).

THEOREM: *A flow pattern* $\mathbf{h}$ *is a system-optimized pattern if and only if for every commodity k there exists an ordering* $1, 2, \ldots, p, p+1, \ldots, m(k)$ *of the chains from* $O^{(k)}$ *to* $D^{(k)}$ *such that:*

$$\begin{aligned} D_1^{(k)}(\mathbf{h}) = D_2^{(k)}(\mathbf{h}) = \cdots D_p^{(k)}(\mathbf{h}) \leqslant D_{p+1}^{(k)}(\mathbf{h}) \leqslant \cdots D_{m(k)}^{(k)}(\mathbf{h}) \\ h_1^{(k)} > 0; \quad h_2^{(k)} > 0; \quad \cdots h_p^{(k)} > 0; \quad h_{p+1}^{(k)} = 0 \cdots h_{m(k)}^{(k)} = 0. \end{aligned} \tag{22}$$

This condition implies that it is possible to obtain a subset of chains with positive flow (possibly only one chain) having equal values of marginal chain cost, and chains carrying zero flow have marginal costs greater than or equal to chains with positive flow. The inequalities in (22) can be obtained directly from the theory of nonlinear programming and are analogous to the complementary slackness conditions used in Sect. 15. The proof of the theorem which we shall now give has such special structure that it gives considerable insight into the traffic assignment process.

We first define a feasible reassignment of the chain flows by

$$\mathbf{h}' = \mathbf{h} + \boldsymbol{\delta} \geqslant \mathbf{0} \tag{23}$$

where $\boldsymbol{\delta}$ is a vector whose elements $\delta_j^{(k)}$ correspond to the elements $h_j^{(k)}$ of $\mathbf{h}$. While $\delta_j^{(k)}$ is unrestricted in sign the flow values from $O^{(k)}$ to $D^{(k)}$ must remain unchanged at

$$\sum_j h_j'^{(k)} = \sum_j h_j^{(k)} = g^{(k)}. \tag{24}$$

Equation (24) implies that for each commodity we must have

$$\sum_j \delta_j^{(k)} = 0 \tag{25}$$

while (23) requires that

$$\delta_j^{(k)} \geqslant -h_j^{(k)}. \tag{26}$$

We now proceed to prove that the condition stated in the theorem is *sufficient*. From (22), (25), and (26) it follows that

$$\sum_1^p \delta_j^{(k)} + \sum_{p+1}^{m(k)} \delta_j^{(k)} = 0 \tag{27}$$

and that

$$\delta_j^{(k)} \geqslant 0 \qquad \text{for} \quad j = p+1, \ldots, m(k). \tag{28}$$

If f' is the new link flow pattern corresponding to the new chain flow pattern $\mathbf{h}'$ given by (23), the change in network cost is

$$C(\mathbf{h}') - C(\mathbf{h}) = \sum_i \left[f_i' \, c_i(f_i') - f_i \, c_i(f_i) \right]. \tag{29}$$

For a strictly convex function $f_i \, c_i(f_i)$, we have shown (Appendix C) that if $f_i' \neq f_i$, i.e., $f_i' > f_i$ or $f_i' < f_i$, then

$$f_i' \, c_i(f_i') - f_i \, c_i(f_i) > (f_i' - f_i) \, d_i(f_i). \tag{30}$$

Thus, in general,

$$C(\mathbf{h}') - C(\mathbf{h}) \geqslant \sum_i (f_i' - f_i) d_i(f_i). \tag{31}$$

Note that a strict inequality holds when one or more $f_i' \neq f_i$. Because $\mathbf{h}' \neq \mathbf{h}$ may lead to cases where $f_i' \equiv f_i$ for all i we must include the possibility that the right-hand side of (31) is zero and $C(\mathbf{h}') = C(\mathbf{h})$. Since

$$f_i' - f_i = \sum_j \sum_k a_{ij}^{(k)} \delta_j^{(k)}, \tag{32}$$

$$C(\mathbf{h}') - C(\mathbf{h}) \geqslant \sum_i \sum_j \sum_k d_i(f_i) \delta_j^{(k)} a_{ij}^{(k)} = \sum_j \sum_k \delta_j^{(k)} D_j^{(k)}(\mathbf{h}) \tag{33}$$

using (20). Typical terms on the right-hand side of (33) can be written as

$$\sum_j \delta_j^{(k)} D_j^{(k)}(\mathbf{h}) = \sum_1^p \delta_j^{(k)} D_j^{(k)}(\mathbf{h}) + \sum_{p+1}^{m(k)} \delta_j^{(k)} D_j^{(k)}(\mathbf{h}). \tag{34}$$

In the first summation, all $D_j^{(k)}(\mathbf{h})$ are equal to $D_1^{(k)}(\mathbf{h})$ by (22), and hence, using (27),

$$\sum_1^p \delta_j^{(k)} D_j^{(k)}(\mathbf{h}) = -D_1^{(k)}(\mathbf{h}) \sum_{p+1}^{m(k)} \delta_j^{(k)} \tag{35}$$

so that (34) becomes

$$\sum_j \delta_j^{(k)} D_j^{(k)}(\mathbf{h}) = \sum_{p+1}^{m(k)} \delta_j^{(k)} [D_j^{(k)}(\mathbf{h}) - D_1^{(k)}(\mathbf{h})]. \tag{36}$$

By (22) and (28) both factors in each term of the summation on the right are nonnegative so that

$$\sum_j \delta_j^{(k)} D_j^{(k)}(\mathbf{h}) \geqslant 0. \tag{37}$$

Substitution in (33) finally gives

$$C(\mathbf{h}') - C(\mathbf{h}) \geqslant 0. \tag{38}$$

If the condition of the theorem is satisfied *any* feasible reassignment of the link flow pattern increases the network cost; any feasible reassignment of the chain flow pattern may or may not increase the network cost. It is rather important to again point out the effects of the nonuniqueness of chain flows in (2) upon the total cost function $C(\mathbf{h})$ in (4) and (21). (See also Sects. 8(c) and (d), Chap. II.) The analysis we have just provided depends upon the convexity of total link costs in terms of link flows but the network cost on the left-hand side of (21) is explicitly written in terms of $\mathbf{h}$, not $\mathbf{f}$. There is usually more than one $\mathbf{h}$ satisfying (2) when $\mathbf{f}$ is fixed; thus even though the chain flow pattern changes, the link flow pattern may not change. If the link flow pattern changes then the strict inequality in (31) and (38) holds.

To show that the condition of the theorem is *necessary* for the minimization of the total network cost, we proceed as in the proof of the previous theorem by assuming that the condition is *not* satisfied so that there exist two chains for which the marginal cost of one chain carrying positive flow is strictly greater than some other chain which may or may not carry flow. Mathematically, for some k, r, s, we have

$$D_r^{(k)}(\mathbf{h}) > D_s^{(k)}(\mathbf{h}) \qquad \text{for} \quad h_r^{(k)} > 0. \tag{39}$$

We now deduce that the traffic pattern $\mathbf{h}$ is *not* system-optimized. We prove this by considering a feasible reassignment $\Delta > 0$ of flow (in the sense of (5)) from chain r to chain s which results in a change of total network cost equal to

$$C(\mathbf{h}') - C(\mathbf{h}) = \sum_i \left[(f_i + \Delta) c_i(f_i + \Delta) - f_i c_i(f_i) \right] a_{is}^{(k)}$$
$$+ \sum_i \left[(f_i - \Delta) c_i(f_i - \Delta) - f_i c_i(f_i) \right] a_{ir}^{(k)}, \tag{40}$$

where the summation on the index i applies only to those links not common to both chains. As we indicated earlier (Fig. 16.1) the change in flows on such links is equal to the reassignment in chain flows. What we now want to show is that we can pick Δ sufficiently small so that the difference in network cost given by (40) is strictly negative. From (11) in Appendix C we obtain the inequalities

$$(f_i + \Delta) c_i(f_i + \Delta) - f_i c_i(f_i) < d_i(f_i + \Delta)\Delta, \qquad \Delta > 0, \tag{41}$$

$$(f_i - \Delta) c_i(f_i - \Delta) - f_i c_i(f_i) < -d_i(f_i - \Delta)\Delta, \qquad \Delta > 0, \tag{42}$$

which state that the difference in total link cost due to the reassignment of flow is strictly less than the new marginal cost of link flow times the amount of flow reallocated. Summing terms in (41) and (42) over all links on chains r and s in (40) that are not common to both chains yields the inequality

$$C(\mathbf{h}') - C(\mathbf{h}) < \sum_i \left[d_i(f_i + \Delta) a_{is}^{(k)} - d_i(f_i - \Delta) a_{ir}^{(k)} \right] \Delta$$
$$= \left[D_s^{(k)}(\mathbf{h}') - D_r^{(k)}(\mathbf{h}') \right] \Delta. \tag{43}$$

But if the inequality of (39) holds for the flow pattern $\mathbf{h}$ we can always pick a Δ sufficiently small and hence a new flow pattern $\mathbf{h}'$ sufficiently close to $\mathbf{h}$ so that

$$D_s^{(k)}(\mathbf{h}') < D_r^{(k)}(\mathbf{h}'), \tag{44}$$

since the marginal cost functions are nondecreasing. Since the term in square brackets in (43) is strictly negative,

$$C(\mathbf{h}') - C(\mathbf{h}) < 0, \tag{45}$$

so that the traffic pattern $\mathbf{h}$ does not minimize the network cost. This completes the proof of the theorem.

(d) Associated Traffic Assignment Problems

Surprising as it may seem, both theorems have strikingly similar mathematical structure. The condition of (12) is identical to that of (22) if one replaces $C_j^{(k)}(\mathbf{h})$ by $D_j^{(k)}(\mathbf{h})$ and preserves the ordering of inequalities on costs and chain flows. The only substantive difference appears to be that we have substituted *marginal* costs of chain flow for *average* costs of a chain. One other observation that may be appropriate at this point is the surprising feature that while one theorem has resulted from strictly local conditions, i.e., comparisons of neighboring routes, the second has been the result of seeking traffic flow patterns which achieve global conditions, i.e., a minimum value for a scalar cost function defined on the entire network.

The two theorems enable us to show an intimate relationship between user-optimized and system-optimized traffic patterns. We show that for any given transportation network with given link costs, $c_i(f_i)$, there is an associated problem with related link costs, $\bar{c}_i(f_i)$, such that the system-optimized pattern for the associated problem is a user-optimized pattern for the given network. In fact, we only have to define

$$\bar{c}_i(f_i) = \frac{1}{f_i} \int_0^{f_i} c_i(x)\,dx. \tag{46}$$

It follows immediately that the marginal link costs for the associated problem are

$$\bar{d}_i(f_i) = \frac{d}{df_i}\left[f_i\bar{c}_i(f_i)\right] = c_i(f_i) \tag{47}$$

and hence the marginal chain costs are

$$\bar{D}_j^{(k)}(\mathbf{h}) = C_j^{(k)}(\mathbf{h}).\qquad(48)$$

By the second theorem, a *necessary* condition for $\mathbf{h}$ to be a system-optimized pattern for the associated problem is given by (22) with $\bar{D}$ instead of D throughout. By (48), this condition becomes precisely that of (12) which is a *sufficient* condition for $\mathbf{h}$ to be a user-optimized pattern of the given network.

The consequences of this result for these associated assignment problems are important. We see that we can determine a user-optimized pattern for a given network by solving a minimum cost pattern for an associated problem on the same network.

(e) A Numerical Example with Four Commodities

An example (see [17]) of the user-optimized and system-optimized flow patterns which are obtained for networks with flow dependent link costs is illustrated in the six-node, ten-link network of Fig. 16.2 having the O–D trip table:

		Destinations	
	1	2	3
$[g^{(k)}]$ = origins 1	0	3	6
3	2	5	0

$$\qquad(49)$$

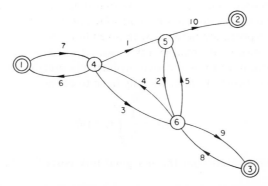

Figure 16.2.   Network with three centroids, three intermediate nodes and ten links. Link numbers are given alongside the links. Link costs of links 1, ..., 5 are flow dependent and link costs of links 6, ..., 10 are zero.

The link costs are given as

$$c_i(f_i) = \frac{1}{5-f_i}, \quad 0 \leqslant f_i < 5, \quad i = 1, ..., 5$$

$$c_i(f_i) = 0, \qquad\qquad i = 6, ..., 10 \tag{50}$$

We number the O–D pairs as follows:

$O^{(1)}$–$D^{(1)}$: trips from node 1 to node 2,

$O^{(2)}$–$D^{(2)}$: trips from node 1 to node 3,

$O^{(3)}$–$D^{(3)}$: trips from node 3 to node 1,

$O^{(4)}$–$D^{(4)}$: trips from node 3 to node 2,

so that our notation for O–D travel demand becomes

$$g^{(1)} = 3; \quad g^{(2)} = 6; \quad g^{(3)} = 2; \quad g^{(4)} = 5. \tag{51}$$

The seven chains that correspond to these four commodities are obtained by listing the distinct links within each set:

$$M^{(1)} = \{(7,1,10); (7,3,5,10)\} \quad \text{with flows} \quad h_1^{(1)}, h_2^{(1)} \tag{52}$$
$$M^{(2)} = \{(7,1,2,9); (7,3,9)\} \quad \text{with flows} \quad h_1^{(2)}, h_2^{(2)} \tag{53}$$
$$M^{(3)} = \{(8,4,6)\} \quad\quad\quad \text{with flow} \quad h_1^{(3)} \tag{54}$$
$$M^{(4)} = \{(8,4,1,10); (8,5,10)\} \quad \text{with flows} \quad h_1^{(4)}, h_2^{(4)} \tag{55}$$

It is interesting to note that link 1 is common to three chains, links 3, 4, and 5 each common to two distinct chains, and link 2 to only one chain. The chain flows must satisfy the conservation equations

$$
\begin{aligned}
h_1^{(1)} + h_2^{(1)} &= 3, \\
h_1^{(2)} + h_2^{(2)} &= 6, \\
h_1^{(3)} \phantom{+ h_2^{(3)}} &= 2, \\
h_1^{(4)} + h_2^{(4)} &= 5.
\end{aligned}
\qquad h_j^{(k)} \geqslant 0, \tag{56}
$$

We see from (52)–(55) that the chain to link flow equations for links $1, ..., 5$ can be written as

$$
\begin{aligned}
h_1^{(1)} + h_1^{(2)} + h_1^{(4)} &= f_1, \\
h_1^{(2)} &= f_2, \\
h_2^{(1)} + h_2^{(2)} &= f_3, \\
h_1^{(3)} + h_1^{(4)} &= f_4, \\
h_2^{(1)} + h_2^{(4)} &= f_5.
\end{aligned}
\tag{57}
$$

First we obtain the user-optimized flow pattern. To do this we consider the associated problem with link costs calculated from (46) to be

$$\bar{c}_i(f_i) = \frac{1}{f_i} \int_0^{f_i} \frac{1}{5-x}\, dx = \frac{1}{f_i} \ln \frac{5}{5-f_i}, \qquad i = 1, \ldots, 5 \tag{58}$$

$$\bar{c}_i(f_i) = 0, \qquad\qquad\qquad\qquad\qquad i = 6, \ldots, 10. \tag{59}$$

The total network cost for the associated problem is therefore

$$\bar{C} = \sum_1^5 \ln \frac{5}{5-f_i} = \ln \prod_1^5 \frac{5}{5-f_i}. \tag{60}$$

To obtain the user-optimized flow pattern we minimize $\bar{C}$ subject to (56) and (57). The required solution may be obtained in this simple case by straightforward differentiation or more generally by a nonlinear programming technique but the method is of no concern here. The user-optimized flow pattern is given in Table 16.1. The optimal network

TABLE 16.1 THE USER-OPTIMIZED FLOW PATTERN

Optimal chain flows	Chain costs	Optimal link flows	Link costs
$h_1^{(1)} = 3.000$	$C_1^{(1)} = 2.794$	$f_1 = 4.642$	$c_1 = 2.793$
$h_2^{(1)} = 0$	$C_2^{(1)} = 6.232$	$f_2 = 1.326$	$c_2 = 0.272$
$h_1^{(2)} = 1.326$	$C_1^{(2)} = 3.066$		
		$f_3 = 4.674$	$c_3 = 3.067$
$h_2^{(2)} = 4.674$	$C_2^{(2)} = 3.066$		
$h_1^{(3)} = 2.000$	$C_1^{(3)} = 0.373$	$f_4 = 2.316$	$c_4 = 0.373$
$h_1^{(4)} = 0.316$	$C_1^{(4)} = 3.166$		
		$f_5 = 4.684$	$c_5 = 3.165$
$h_2^{(4)} = 4.684$	$C_2^{(4)} = 3.166$		

cost is $C = 43.351$. It can be verified that the conditions of the theorem proved in Sect. 16(b) are satisfied. For example,

$$C_1^{(1)} < C_2^{(1)} \qquad \text{with} \quad h_1^{(1)} > 0, \quad h_2^{(1)} = 0 \tag{61}$$

and

$$C_1^{(2)} = C_2^{(2)} \qquad \text{with} \quad h_1^{(2)} > 0, \quad h_2^{(2)} > 0. \tag{62}$$

Secondly we obtain the system-optimized flow pattern by minimizing the network cost, given by

$$C = \sum_1^5 \frac{f_i}{5-f_i}, \tag{63}$$

subject to (56) and (57). The system-optimized flow pattern is given in Table 16.2. The optimal network cost is $C = 43.238$. Again we can

TABLE 16.2 THE SYSTEM-OPTIMIZED FLOW PATTERN

Optimal chain flows	Marginal chain costs	Optimal link flows	Link costs
$h_1^{(1)} = 3.000$	$D_1^{(1)} = 44.643$	$f_1 = 4.665$	$c_1 = 2.988$
$h_2^{(1)} = 0$	$D_2^{(1)} = 90.361$	$f_2 = 1.333$	$c_2 = 0.273$
$h_1^{(2)} = 1.333$	$D_1^{(2)} = 45.015$	$f_3 = 4.667$	$c_3 = 3.001$
$h_2^{(2)} = 4.667$	$D_2^{(2)} = 45.015$	$f_4 = 2.332$	$c_4 = 0.375$
$h_1^{(3)} = 2.000$	$D_1^{(3)} = 0.702$	$f_5 = 4.668$	$c_5 = 3.012$
$h_1^{(4)} = 0.332$	$D_1^{(4)} = 45.346$		
$h_2^{(4)} = 4.668$	$D_2^{(4)} = 45.346$		

verify the conditions of the theorem proved in Sect. 16(c). The marginal link costs are

$$d_i(f_i) = \frac{5}{(5-f_i)^2}, \qquad i = 1, ..., 5 \tag{64}$$

and from the table we see that the marginal chain costs satisfy, for example,

$$D_1^{(1)} < D_2^{(1)} \qquad \text{with} \quad h_1^{(1)} > 0, \quad h_2^{(1)} = 0 \tag{65}$$

and

$$D_1^{(2)} = D_2^{(2)} \qquad \text{with} \quad h_1^{(2)} > 0, \quad h_2^{(2)} > 0. \tag{66}$$

To see just how complicated the interactions between chain flows and chain costs really are, consider the effect of a small decrease in the chain flow $h_1^{(4)}$. $h_2^{(4)}$ must be increased to preserve feasibility. This reassignment decreases f_1 and f_4 and increases f_5 because these links are incident upon both chains 1 and 2 of commodity 4. Obviously c_1 and c_4 must decrease while c_5 increases. These changes then force $C_1^{(1)}$, $C_1^{(2)}$, $C_1^{(3)}$, and

$C_1^{(4)}$ to decrease while $C_2^{(1)}$ and $C_2^{(4)}$ must increase. In other words, the reassignment of one chain flow affects the costs of six out of the seven chains that we have considered thus far!

(f) Congested Assignment

Despite the elegance of the theoretical analysis of extremal principles for flow-dependent costs, it has not led to assignment procedures or programs which can handle large transportation networks. Many *congested assignment* programs, as they are called, have been used in transportation planning but the present state of the art is such that no one program has been widely accepted.

Most congested assignment procedures are iterative. For certain link costs, traffic is assigned and then the link costs recalculated, traffic reassigned, and so on. It is a pious hope that this iterative procedure should converge, and although possible oscillations can be overcome or damped out the resultant assignments become less meaningful.

For the special case when the link costs are step functions, it is possible to formulate the congested assignment as a linear program which can cope with reasonably large networks. For example, Charnes and Cooper [24] have formulated a multicopy traffic assignment model which, in the link flow notation of Sect. 8(d), Chap. II, can be expressed as the following mathematical program:

$$\mathbf{f}^{\alpha} \geqslant \mathbf{0}, \tag{67}$$

$$\mathbf{Ef}^{\alpha} = \mathbf{g}^{\alpha}, \tag{68}$$

$$\mathbf{f} = \sum_{\alpha} \mathbf{f}^{\alpha} \leqslant \mathbf{u}, \tag{69}$$

$$\mathbf{c(f)}^{\mathrm{T}} \mathbf{f} = C(\min), \tag{70}$$

where $\mathbf{f}^{\alpha}$ is the unknown $l \times 1$ link flow vector with elements equal to the link flow of copy α; $\mathbf{g}^{\alpha}$ is the given $n \times 1$ copy flow vector; $\mathbf{E}$ is the $n \times l$ node–link incidence matrix; $\mathbf{u}$ the $n \times 1$ vector of link capacities; and $\mathbf{c(f)}$ the $l \times 1$ vector of link costs. The increase in link cost with flow is represented by a step function with a separate parallel capacitated link for each step. For example, two parallel links $i = 1, 2$, connecting the same nodes, one with link cost c_1 and capacity u_1 and the other with link cost c_2 and capacity u_2, and $c_2 > c_1$, give the piece-wise linear function

$$fc(f) = \begin{cases} c_1 f, & 0 \leqslant f \leqslant u_1, \\ c_1 u_1 + c_2(f - u_1), & u_1 \leqslant f \leqslant u_2. \end{cases} \tag{71}$$

Thus, by coding the network with suitably chosen parallel links where necessary, the multicopy assignment is still expressed as a linear program and can be solved by the usual LP algorithms. Charnes and Cooper illustrate the theory with a multicopy assignment to a street network with 18 nodes, 27 links, and 11 origins giving 11 copies. A more detailed example has been given by Pinnell and Satterly [25].

Tomlin [26] has explicitly formulated congested assignment as a minimum network cost multicommodity flow problem. With the notation of Sect. 8(d), Chap. II, the problem can be expressed as the following mathematical program:

$$h_j^{(k)} \geqslant 0, \tag{72}$$

$$\sum_j h_j^{(k)} = g^{(k)}, \tag{73}$$

$$f_i = \sum_k \sum_j a_{ij}^{(k)} h_j^{(k)} \leqslant u_i, \tag{74}$$

$$\sum_k \sum_j C_j^{(k)} h_j^{(k)} = C(\min), \tag{75}$$

where $C_j^{(k)}$ is the route cost for the chain $m_j^{(k)}$. If the costs are step functions, this program is a linear program and can be solved by the multicommodity flow algorithms. Unfortunately these are inefficient, and the complete enumeration of the O–D chains for a large network would be impossible. This may be overcome by a column generating technique which introduces new chains calculated by a cheapest route algorithm. Use is made of the dual linear program:

$$v^{(k)} \quad \text{unrestricted in sign,} \tag{76}$$

$$\mu_i \geqslant 0, \tag{77}$$

$$v^{(k)} - \sum_{(j)} \mu_i \leqslant C_j^{(k)}, \tag{78}$$

$$\sum v^{(k)} g^{(k)} - \sum \mu_i u_i = V(\max). \tag{79}$$

In (78) the dual variables μ_i are, by virtue of (74), independent of the commodity k; the summation in this equation follows the convention adopted in (43), Sect. 15. Various techniques have been tried in an attempt to improve the efficiency [27], but practical applications have in the main been restricted to small networks.

17. Notes and References

[1] Wardrop, J. G., Some Theoretical Aspects of Road Traffic Research, *Proc. Inst. Civil Eng.*, **Part II**, 325–378 (1952).

This interesting paper and the ensuing printed discussion summarizes some of the early research work done at the Road Research Laboratory, England, which has pioneered and stimulated the dramatic recent growth of traffic research throughout the world. The paper emphasizes the value of a theoretical approach to traffic problems and illustrates this with an analysis of the following: flow and speed of traffic, frequency of overtaking, capacity of road systems, signal-controlled intersections, formation of queues, distribution of traffic over alternative routes, before-and-after studies. Although the paper was published some years ago, it still serves as an excellent introduction to traffic theory.

It is curious that as often as not the first extremal principle enunciated by Wardrop is referred to in the literature as his second, and the second as his first!

[2] Murchland, J. D., Bibliography of the Shortest Route Problem, *London Business Studies Report*, LBS-TNT-6.2 (1969).

The reports from the short-lived Transport Network Theory Unit of the London School of Economics and Political Science (and subsequently of the London Graduate School of Business Studies) form a very significant and lively contribution to transportation science, and it is regretable that few of the reports have found their way into well-read journals. Report No. 6.2 is a second revision of an extensive bibliography, containing 97 references.

Murchland classifies the shortest route algorithms into three main categories—tree building, matrix, and partitioned.

[3] Dreyfus, S. E., An Appraisal of Some Shortest-path Algorithms, *Operations Res.*, **17**, 395–412 (1969).

The purpose and scope of this interesting paper can be best summarized by quoting first from the author's introduction and then from his conclusion: "... it is hoped that our somewhat skeptical survey of current literature will put the interested reader on guard and perhaps save him, or his digital computer, considerable time and trouble;" and "Our interest has not been definitive solution, but rather to clear the air by presenting both some important methods and references and some critical comments and warnings."

[4] Kirby, R. F. and Potts, R. B., The Minimum Route Problem

for Networks with Turn Penalties and Prohibitions, *Transportation Res.*, **3**, 397–408 (1969).

This paper surveys the literature and resurrects an important paper which had been apparently overlooked. It also points out an erroneous procedure which has been suggested in the literature.

[5] Bellman, R., On a Routing Problem, *Quart. Appl. Math.*, **XVI**, 87–90 (1958).

This comparatively early paper is important because it is one of the few in the literature which gives a precise statement and proof of a minimum route algorithm. The present text uses the dynamic programming functional equation technique described in this paper. An earlier and similar analysis of the shortest route problem is found on pp. 52–54 of the monograph by Beckmann [6], Sect. 12, Chap. II.

[6] Dijkstra, E. W., A Note on Two Problems in Connexion with Graphs, *Numer. Math.*, **1**, 269–271 (1959).

This short paper briefly describes the algorithm which is analyzed in detail in the text. We have purposely avoided attaching authors' names to algorithms because of the general confusion in the literature on cheapest routes.

[7] Caldwell, T., On Finding Minimum Routes in a Network with Turn Penalties, *Comm. ACM*, **4**, 107–108 (1961).

This important paper seems to have been overlooked and has, for example, escaped inclusion in Murchland's 1967 bibliography. Perhaps one reason for its neglect is that it does not formulate a practical algorithm directly applicable to networks as they are usually coded in transportation planning.

[8] Brokke, G. E., Urban Transportation Planning Computer System, *American Association State Highway Officials Conference*, Minnesota (1967).

The Bureau of Public Roads has cooperated with State Highways Departments for many years in establishing transportation planning surveys. Their computer packages developed in the late 1950s for use on first generation computers such as the IBM 704 have been widely used and have been largely responsible for defining the analytic computer oriented approach to the planning process. Early in the 1960s, the computer packages were revised for the second-generation computers (e.g., IBM 7094) and the Traffic Assignment Manual (1964) is one of the excellent publications which explain in detail the use of the programs. The Bureau is currently rewriting the programs for the third generation computers (IBM 360), and the new tree or vine building programs such as described in this reference include an option which permits turn penalties and prohibitions to be correctly counted.

[9]   Control Data Corporation Data Centers Division, Users'
      Manual, Transportation Planning System for the Control
      Data 3600 Computer (1965).
Control Data Corporation has developed a transportation planning system, called TRANPLAN, which has been widely used as an alternative to the BPR program packages, although basically the computer programs are very similar. The system is being rewritten for the 6000 series of Control Data computers.

[10]   Sema, Group Metra France, Le Modele ATCODE (1964).
Metra has provided a significant contribution to transportation planning by its development of a comprehensive package of computer programs, including generation, distribution, modal split, and assignment.

[11]   Kirby, R. F., A Minimum Path Algorithm for a Road Network
       with Turn Penalties, *Proceedings of the Third Australian Road
       Research Board Conference, Part I*, 434–442 (1966).
A fully documented version of this algorithm has been lodged with the National Association of Australian State Road Authorities and has been used with success in several metropolitan transportation studies. The program has been written for a 32K Control Data 3600 Computer for networks of up to 3000 nodes (including a maximum of 650 centroids) and 12,000 links with 32 turn penalty types available for each turn. The program has been modified for use on a Control Data 6400 Computer and includes a facility for specifying intersection types as well as turn types.

[12]   Wachs, M., Relationships Between Drivers' Attitudes Toward
       Alternate Routes and Driver and Route Characteristics, *H.R.B.
       Record*, **197**, 70–87 (1967).
This paper reports a home-interview study designed to analyze the factors which drivers consider important in the choice of routes for different purposes—work trip, shopping, and a trip to visit a friend. The drivers expressed their preferences for access controlled routes, shortest routes, safety, congestion, strain, pleasant scenery, etc. A careful statistical analysis led to the conclusion that reasonably strong relationships do exist between the attitudes of drivers toward the type of

route they seek when they make a trip, and the characteristics of the drivers, their trips, and the routes to which they have been exposed.

[13]   Jansen, G. R., A Pilot Study in Trip Assignment, *I.T.T.E. Graduate Report*, University of California, Berkeley (1966).

This report gives the result of a survey made of drivers' choices of routes between the Plaza of the Golden Gate Bridge and the downtown office of the California State Automobile Association. The travel times on roads of the network were measured by the floating car method, and the results differed quite markedly from those incorporated in the BATS networks. From travel diaries kept by 17 drivers, Jansen found that five significantly different routes were chosen and of these the most popular (in the morning peak) was the next-to-longest in travel time! Turn penalties and prohibitions were not taken into account.

[14]   Stover, V. G., The Texas Large Systems Traffic Assignment Package, *Traffic Quart.*, **21**, 339–354 (1967).

This paper describes a battery of programs TEXAS–BIGSYS which can carry out cheapest route traffic assignments for a main road network with 16,000 nodes (including 4800 centroids) and 64,000 links. This may be a significant computing achievement but hardly a useful tool for the planner.

[15]   Michaels, R. M., Attitudes of Drivers Determine Choice between Alternate Highways, *Public Roads*, **33**, 225–236 (1965).

This paper reports a study carried out on driver's choice between two routes—one the Maine Turnpike from Kittering to South Portland, U.S.A., and the other the parallel rural highway US 1. Over three thousand drivers were questioned, and from their answers the author concluded that drivers did have stable attitudes which correlated with their choices. Test drivers were used to measure the tension or stress characteristics of the alternate routes and these were shown to be a significant factor in determining the diversion to the freeway route.

[16]   Burrell, J. E., Multiple Route Assignment and its Application to Capacity Restraint, *Fourth International Symposium on Theory of Traffic Flow*, Karlsruhe (1968).

This interesting paper reports on a novel assignment program developed in conjunction with the London Transport Study, England. The rectangular link cost probability function is characterized by the mean absolute deviation, its quotient with the mean link cost being taken the same for all links. The possibility of choosing one of eight link costs still gives a unique O–D route, but within the network many links will be used thus spreading out the traffic flows.

[17]   Jorgenson, N. O., Some Aspects of the Urban Traffic Assignment Problem, *I.T.T.E. Graduate Report*, University of California, Berkeley (1963).

It is unfortunate that some of the material in this report did not appear earlier in the open literature. Partly because of this, we have given Jorgenson's treatment of the link flow linear programming approach to the two extremal principles in detail, although his statement of the extended first principle has been somewhat modified. His derivation of the equivalent mathematical problems for user-optimized and minimum network cost traffic patterns with convex link costs predates other published formulations of the model. In the context of transportation problems, he showed that the flow versus travel time relationships were identical in structure to the Kuhn–Tucker optimality conditions of an equivalent extremal problem. While the author is primarily concerned with a link flow formulation, we have chosen to emphasize the chain flow aspects of traffic assignment.

[18]   Dantzig, G. B., *Linear Programming and Extensions*, Princeton Univ. Press, Princeton, New Jersey (1963).

Of the many texts on linear programming, we have chosen to refer to this one because it is authoritative, readable, and contains chapters on network flows and the cheapest route and minimum network cost problems. In Chap. 17, the author emphasizes the importance of the relationship between trees and basic solutions. In a lucid paper (*SIAM Rev.*, **10**, 371–372 (1968)), A. F. Veinott and G. B. Dantzig give a concise proof that the total unimodularity of a matrix $\mathbf{A}$ is a necessary and sufficient condition for integral inverses of every basis of the system of inequalities $\mathbf{x} \geqslant \mathbf{0}$, $\mathbf{Ax} \leqslant \mathbf{b}$, where $\mathbf{A}$ and $\mathbf{b}$ are integral. The node–link incidence matrix $\mathbf{E}$ in (4) (Sect. 15), the augmented matrix $(\mathbf{E}, \mathbf{I})$ implicit in (20) and (21) (Sect. 15), the link–chain incidence matrix $\mathbf{A}$ implicit in (39) (Sect. 15) are examples of totally unimodular matrices. Thus, the linear programs of Section 15(b) yield integral basic solutions for link and chain flows when the given flow values and link capacities are integers. The property of total unimodularity does not extend to the class of augmented incidence matrices that characterize multicommodity and multicopy flows.

[19]   SHARE Program SDA 3536, Out-of-Kilter Network Routine, OKF3 (1967).

This program solves the minimum network cost single commodity problem for a capacitated network with given constant link costs. The abstract for this well-documented program is as follows: "An independent routine to solve capacitated network flow problems using a method in which a measure of optimality is not worsened on any iteration. Flows have upper and lower bounds which may be positive or negative. No initial feasible solution is needed. Has provision for solving problems which vary slightly from previously solved problems in minimal machine time. Source language is Fortran IV." This program and variants of it have been widely used.

It is interesting to note that our simplified presentation of the out-of-kilter algorithm, which assumes an initial feasible solution, must have been similar to that first drafted by Fulkerson. In his original paper— An Out-of-Kilter Method for Minimal-cost Flow Problems, *SIAM J.*, **9**, 18–27 (1961), which is the basis for the relevant section in the Ford and Fulkerson text ([1], Sect. 11, Chap. II)—Fulkerson expresses his appreciation to Dantzig "whose criticism of an earlier version of this paper in which the initial x was assumed feasible, led us to reconsider the problem from the standpoint of infeasible x."

[20]   Nash, J., Non-Cooperative Games, *Ann. of Math.*, **54**, No. 2, (September 1951).

This highly original paper develops the notion of an equilibrium point for n-person games in which players do not cooperate or form coalitions. The author shows that every such game has at least one equilibrium point. The analogy with traffic assignment on a transportation network is that each commodity represents a player; a traffic flow equilibrium corresponds to the case where travelers having routes with a given O–D pair minimize their travel costs when the routes used by all other commodities are held fixed. The traffic patterns of (11) (Sect. 16) are more properly referred to as *Nash equilibria*. Possibly one of the most important requirements that is implicitly required in Nash's theory (but not necessarily attained in real-life transportation networks) is the availability of complete information on the status of alternate routes.

[21]   Dafermos, Stella C., and Sparrow, F. T., The Traffic Assignment Problem for a General Network, *NBS J. Res. Ser. B*, **73**, 91–118 (1969); Dafermos, Stella C., An Extended Traffic Assignment Model with Applications to Two-way Traffic, *Transportation Sci.*, **5**, 366–389 (1971).

These two papers (the later one more readable than the first) give a rigorous mathematical treatment of traffic assignment problems for flow dependent costs. The careful distinction made between user-optimized and system-optimized traffic patterns has been followed in our text, together with the statements and proofs of the two important theorems we give in Section 16. The authors also develop an interesting algorithm which begins with an initial feasible flow pattern and by means of an "equilibration operator" constructs a sequence of feasible flow patterns which converges to the user-optimized solution. Because of the relationship between the user-optimized and system-optimized solutions as discussed in Sect. 16(d), this algorithm also provides a method for solving the minimum network cost problem.

[22]   Beckmann, M. J., On the Theory of Traffic Flow in Networks, *Traffic Quart.*, **21**, 109–117 (1967).

The author formulates an abstract and general model of flows in transportation networks. Most of the important contributions in this field prior to 1967 can be viewed as special cases of his model. Although his notation differs substantially from our own, our material covers the five classes of relations that specify his transportation network model: (1) average costs as a function of link flow, (2) the relationships between O–D travel demand and the cost (or time) of travel, (3) the relationship between route costs and link costs, (4) the conservation equations which state that link flows are the sum over all commodities of commodity flows on each link, and (5) Kirchhoff's conservation equations for each commodity or copy. He also recognizes the existence of a "suitable" mathematical function whose extremal values allow one to characterize the equilibrium flow patterns in transportation networks. He does not, however, explicitly formulate them in terms of Nash equilibria. See also [6], Sect. 12, Chap. II.

[23]   Kitchen, J. W., *Calculus of One Variable*, Addison-Wesley, Reading, Massachusetts (1968).

Chapter 6 of this book has an excellent and simple two-dimensional treatment of convex functions and convex sets. The mean-value theorems, the three-chords lemma, and some of the inequalities that we use are stated and proved in this very readable text. Although the material can be used as background for an advanced course, this calculus book is often used at the freshman and sophomore college level.

[24]   Charnes, A. and Cooper, W. W., Multicopy Traffic Network Models, in *Theory of Traffic Flow* (R. Herman, Editor), 84–96, Elsevier, Amsterdam (1961).

This paper was presented at the First International Symposium on the Theory of Traffic Flow held at G. M. Research Laboratories, Michigan, U.S.A. The theory is illustrated with an interesting example computed from a modification of data pertaining to a small town in Indiana.

[25] Pinnell, C. and Satterly, G. T., Analytical Methods in Transportation: Systems Analysis for Arterial Street Operations, *J. Engrg. Mech.—Div., Proc. Amer. Soc. Civ. Eng.*, **89**, 67–95 (1963).

The title of this paper perhaps disguises the content, which is a detailed example of a multicopy assignment to a network with 25 nodes, 176 links, 12 origins, and 3 destinations. The copy flows were identified as those with particular destinations, giving a total of 3 copies. All except 4 links were taken as uncapacitated, and on these four a step cost function was assumed. The same example was solved using a discrete version of Pontryagin's principle by Funk, M. L., Snell, R. R., and Blackburn, J. B., (see *J. Hway.—Div., Proc. Amer. Soc. Civ. Eng.*, **93**, 95–113 (1967)).

[26] Tomlin, J. A., Minimum Cost Multicommodity Network Flows, *Operations Res.*, **14**, 45–51 (1966).

In this paper, the author gives a node–link and a link–chain formulation of the minimum network cost problem and discusses their equivalence. A more detailed discussion is contained in the author's PhD thesis: Mathematical Programming Models for Traffic Network Problems, University of Adelaide, South Australia, 1967.

[27] Gibert, A., A Method for the Traffic Assignment Problem, *Report LBS-TNT-95*, London Graduate School of Business Studies (1968).

This paper suggests a method for improving the efficiency of a multicommodity assignment program by storing and not discarding cheapest routes determined for nonoptimal solutions. Experience shows that for each commodity flow between a particular O–D pair, few of the possible chains are used, and these are usually determined in the first few interations.

18.   Problems

1. For the undirected network illustrated in Fig. 18.1, use the algorithm described in Sect. 14(b) to find the cheapest route tree with node 1 as

home node. Is the cheapest route from node 1 to node 6 altered if the link costs are all increased by 2 units?

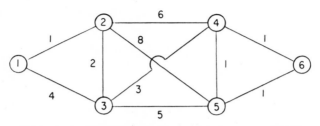

Figure 18.1.   Undirected network with link costs as indicated.

2. Find a second-to-cheapest path from node 1 to node 6 in the network illustrated in Fig. 18.1. Describe a general algorithm for finding second-cheapest routes on a network.

3. Consider an undirected connected network with given link lengths. Is the following statement true: if a *dearest* path from node n_1 to node n_r passes through node n_i, then that portion of the path from n_1 to n_i is a dearest path from n_1 to n_i? Illustrate your answer by considering the network in Fig. 18.1 and taking $n_1 = 1$, $n_i = 2$, $n_r = 3$.

4. For the network $[N; L]$ illustrated in Fig. 18.1 define $c_{ii} = 0$, $i \in N$, and $c_{ij} = \infty$, $(i,j) \notin L$. Compute the following for $i,j = 1, 2, ..., n$:
Step $k = 0$:

$$C_{ij}^{(o)} = c_{ij} ;$$

Steps $k = 1, 2, ..., n$:

$$C_{ij}^{(k)} = \min[C_{ij}^{(k-1)}, C_{ik}^{(k-1)} + C_{kj}^{(k-1)}] .$$

Show that $C_{ij}^{(n)} = C_{ij}^*$ is the cost of a cheapest path from node i to node j, and justify the algorithm.

5. With the same definitions as in Problem 4, compute successively matrices of order $k = 1, 2, ..., n$ with elements $C_{ij}^{(k)}$ as follows:
Step $k = 1$:

$$C_{11}^{(1)} = 0 ;$$

Steps $k = 2, 3, ..., n$: For $i, j = 1, 2, ..., k-1$, define

(a) $C_{ik}^{(k)} = \min_j [C_{ij}^{(k-1)} + c_{jk}]$;

(b) $C_{ki}^{(k)} = \min_j [c_{kj} + C_{ji}^{(k-1)}]$;

(c) $C_{kk}^{(k)} = 0$;

(d) $C_{ij}^{(k)} = \min [C_{ij}^{(k-1)}, C_{ik}^{(k)} + C_{kj}^{(k)}]$.

Show that $C_{ij}^{(n)} = C_{ij}^*$ is the cost of a cheapest path from node i to node j, and justify the algorithm.

6. Consider the following statement for a network without turn penalties in relation to (15), Sect. 14: If $0 < C^*(n_1, n_j) - C^*(n_1, n_i) < c(n_i, n_j)$, then there is no cheapest path from n_1 to n_j which passes through n_i. Is this statement true or false?

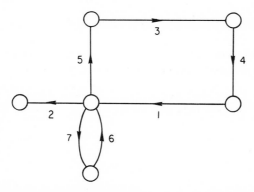

Figure 18.2. Network with prohibited turn. This network is the same as that in Fig. 14.4, except that the links are now numbered $1, 2, ..., 7$.

7. For the network in Fig. 18.2, suppose that the link costs are

$$c(1) = c(2) = c(3) = 20, \qquad c(4) = c(5) = 60,$$

$$c(6) = c(7) = 90,$$

and that the turn penalties are

$$p(1, 2) = p(6, 5) = 0, \qquad p(1, 5) = p(3, 4) = p(4, 1) = p(5, 3) = 5,$$

$$p(1, 7) = 15,$$

with all other values of $p = \infty$. Compute the cheapest route tree with link 6 (or more specifically the beginning of this link) as home link.

8. In an attempt to account for turn penalties, the tree-building algorithm described in Sect. 14(b) is modified as follows: to the condition

$$(n_{k-1}, n_j) \in (X_{k-1}, \overline{X}_{k-1})$$

in (i) for steps $k = 2, 3, \ldots, n$ is added "and the sequence of the three nodes $P(n_{k-1})$, n_{k-1}, n_j corresponds to an allowable turn", etc. Show that this new plausible algorithm is incorrect by applying it to the following network:

$$N = \{1, 2, 3, 4\},$$

$$L = \{(1, 2), (1, 3), (1, 4), (2, 3), (3, 4)\},$$

$$c(1, 2) = 3, \qquad c(1, 3) = 1, \qquad c(1, 4) = 6,$$

$$c(2, 3) = 1, \qquad c(3, 4) = 1,$$

turn from link $(1, 3)$ to link $(3, 4)$ is prohibited.

9. Consider the directed single O–D capacitated network illustrated in Fig. 15.2 with corresponding link capacities and costs as follows:

Link number i	Link capacity u_i	Link cost c_i	Link flow f_i	Dual variable μ_i
1	6	1	5	0
2	8	4	1	0
3	2	2	2	1
4	4	6	3	0
5	1	8	0	0
6	5	3	3	0
7	3	5	0	0
8	4	1	0	0
9	8	1	6	0
10	9	1	0	0

(i) Show that the tabulated link flows f_i give a feasible network flow with flow value $g = 6$ and network cost $C = 46$ units.

(ii) Show that the tabulated dual variables μ_i give a feasible solution of the dual LP with $v = 8$ and $V = 46$ units.

(iii) Verify the complementary slackness relations (47)–(50) (Sect. 15).

(iv) Interpret the optimal chain flow solution in relation to the extended form of the first extremal principle.

10. Repeat Problem 9 using the same network and the same link capacities and costs but take

$$g = 11, \quad \text{and} \quad \mathbf{f}^T = [6\,5\,2\,4\,0\,5\,2\,1\,8\,3],$$

$$v = 10, \quad \text{and} \quad \boldsymbol{\mu}^T = [1\,0\,0\,0\,0\,1\,0\,0\,1\,0].$$

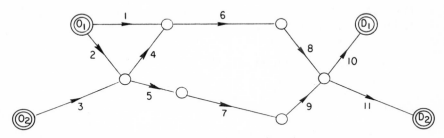

Figure 18.3. Multiple O–D network for traffic assignment. O_1 and O_2 are the origins, D_1 and D_2 are the destinations, and the links are numbered $1, 2, \ldots, 11$.

11. Figure 18.3 represents a directed network with 2 origins O_1, O_2, 2 destinations D_1, D_2, 6 intermediate nodes, and 11 links numbered $i = 1, 2, \ldots, 11$.

(i) Enumerate all chains from the origins to the destinations and determine their route costs if the link costs are given by

Link: $\quad i \quad$ 1 2 3 4 5 6 7 8 9 10 11

Link cost: $\quad c_i \quad$ 3 1 5 1 1 2 4 1 1 1 3 .

(ii) The O–D flows are given by the trip table

$$
\begin{array}{c@{\quad}cc@{\quad}c}
 & D_1 & D_2 & \\
O_1 & \begin{bmatrix} 35 & 20 \end{bmatrix} & & 55 \\
O_2 & \begin{bmatrix} 15 & 30 \end{bmatrix} & & 45 \\
 & 50 & 50 & 100 = v.
\end{array}
$$

Assuming that the links are uncapacitated, assign these trips and determine the commodity and link flows using all-or-nothing cheapest route assignment. What is the network cost?

12. With the same data as in Problem 11, determine the cheapest routes

to the destinations from origins O_1 and O_2, and find the corresponding copy and link flows using all-or-nothing cheapest route assignment.

13. Repeat Problem 12 using copy flows to D_1 and D_2.

14. With the same data as in Problem 11, suppose that the link $i = 4$ has a capacity $u_4 = 50$ units, all other links remaining uncapacitated. Using all-or-nothing cheapest route assignment, assign the O–D trips to the network

    (i)   in the order O_1–D_1, O_1–D_2, O_2–D_1, O_2–D_2,

    (ii) in the order O_2–D_2, O_2–D_1, O_1–D_2, O_1–D_1.

Determine the link flows and network cost in each case.

15.   With the same data as in Problem 14, determine the network flow which minimizes the network cost.

16. Links 6 and 7 in the network in Fig. 18.3 have link costs given by

$$c_6(f) = \begin{cases} 1, & 0 \leqslant f \leqslant 20, \\ 1 + 0.4(f - 20), & f \geqslant 20, \end{cases}$$

$$c_7(f) = \begin{cases} 3, & 0 \leqslant f \leqslant 30, \\ 3 + 0.1(f - 30), & f \geqslant 30. \end{cases}$$

All other links have constant link costs

Link:	i	1	2	3	4	5	8	9	10	11
Link cost:	c_i	3	3	5	5	1	1	3	1	3

For the trip table as in Problem 11, assign the O–D trips to give a user-optimized flow pattern. Are the optimal chain flow and link flow patterns unique? What is the network cost?

17. For the same data as in Problem 16, find a system-optimized flow pattern. Are the optimal chain flow and link flow patterns unique? What is the minimum network cost?

18. Consider a single O–D capacitated network $[N; L]$ with origin node 1 and destination node n. By introducing an uncapacitated return link $(n, 1)$, formulate the maximal flow problem as a linear program, both primal and dual, and prove the max-flow min-cut theorem.

TRIP DISTRIBUTION

19. Introduction

Trip distribution plays an important role in the analysis and network evaluation phases of the transportation planning process. The concept of trip or trip interchange is somewhat loosely and ambiguously defined. Usually it refers to interzonal journeys from one zone (origin) to another zone (destination) and to intrazonal journeys within a zone. Sometimes the trips refer to total person trips, sometimes to vehicle trips. The number of trips is estimated for a particular period of time and may refer, for example, to an average weekday or peak hour. The trip numbers, therefore, have the dimensions of flow and the two terms will be used interchangeably. The trips are usually classified according to purpose, the main stratifications being home-based work trips, home-based nonwork trips and nonhome-based trips. This awkward terminology is somewhat self-explanatory. Special consideration may be given to external trips, i.e., trips which have origin or destination outside the study area.

The essential problem in trip distribution is to determine, from the estimated number of trips produced at and attracted to each zone, the number of interzonal and intrazonal trips. The numbers of trips can be regarded as elements of a trip table or distribution matrix or as flows on a traffic desire network.

In the planning process, trip distributions are evaluated in conjunction with traffic assignment models, such as have been described in the previous chapter. Most trip distribution models use as parameters the interzonal travel times or costs but in general these depend on the network and the traffic assigned to it. Thus, the output of the traffic assignment program is required as an input to the trip distribution program. On the other hand, the output of the trip distribution program, giving the interzonal trips, forms an essential input to the traffic assignment program. This feedback between traffic assignment and trip distribution is especially important in the evaluation of new networks.

We shall in this chapter first formulate trip distribution as a mathematical model and then analyze in some detail a variety of these models. The first we shall discuss is the Hitchcock model [1] which in the literature on network flows is usually associated with *the* transportation problem. This model features both extremal principles which were discussed in the previous chapter. Next we proceed to analyze a class of distribution models which are based on the concept of network entropy. Included in this class is the gravity model, which has served as one of the mainstays of transportation planning. In the following section we give a brief description of opportunity models, with special attention to a preferencing model. In order to interrelate the concepts of this chapter and the previous one, we conclude with some models which simultaneously consider trip distribution and traffic assignment.

20. Model Formulation

Formulated as a mathematical problem, trip distribution requires the determination of the number of trips f_{ij} from centroid i to centroid j given the total number of trips a_i produced at i, the total number of trips b_j attracted to j, and the cost c_{ij} of a trip from i to j. The essence of the mathematical model is the functional dependence of f_{ij} on the a_i, b_j, and c_{ij}.

As usual, cost may be interpreted as travel time, distance or a combination of these. The c_{ij} are often determined in the transportation planning programs as output of a "skim trees" routine giving the travel times of the shortest (i.e., quickest) interzonal routes. The quantities c_{ii} are chosen as mean travel times for intrazonal trips.

There are various constraints which may be regarded as desirable properties of a trip distribution model. For example, one might require that the f_{ij} satisfy the conservation laws

$$\sum_j f_{ij} = a_i, \tag{1}$$

$$\sum_i f_{ij} = b_j, \tag{2}$$

with the understanding that

$$\sum_i a_i = \sum_j b_j = v, \tag{3}$$

where v is the total number of trips or flow value. Although this seems an obvious mathematical requirement, some models do not force exact equality in (1) and (2), and there is some justification for this in transportation planning, because the values of a_i and b_j are usually not known with any great accuracy.

In addition to conservation, the model may be required to satisfy the compressibility and separability constraints described in Sect. 8(e), Chap. II. For example, if centroids $n-1$ and n of a traffic desire network with n centroids are combined, compressibility requires the new trip numbers f_{ij}', productions a_i', attractions b_j', and total trips v' to be related to the old by the equations

$$f_{ij}' = f_{ij}, \qquad\qquad a_i' = a_i, \qquad b_j' = b_j, \tag{4}$$

$$f_{n-1,j}' = f_{n-1,j} + f_{nj}, \qquad\qquad a_{n-1}' = a_{n-1} + a_n, \tag{5}$$

$$f_{i,n-1}' = f_{i,n-1} + f_{in}, \qquad\qquad b_{n-1}' = b_{n-1} + b_n, \tag{6}$$

$$f_{n-1,n-1}' = f_{n-1,n-1} + f_{n-1,n} + f_{n,n-1} + f_{nn}, \tag{7}$$

for $i,j = 1, 2, ..., n-2$. Note that

$$v' = \sum a_i' = \sum a_i = v. \tag{8}$$

Separability requires that if node n were removed, then the new trip numbers, productions, attractions, and total trips should be

$$f_{ij}' = f_{ij}, \tag{9}$$

$$a_i' = a_i - f_{in}, \tag{10}$$

$$b_j' = b_j - f_{nj}, \tag{11}$$

$$v' = v - a_n - b_n, \tag{12}$$

for $i,j = 1, 2, ..., n-1$.

Both compressibility and separability are desirable properties of a trip distribution model, because they imply some independence of the particular manner in which the study area has been subdivided into zones.

Finally we may wish to force the model to give nonnegative integral values of f_{ij}. In some models this is automatically satisfied; in others it is an important constraint.

In practice it is not possible or even desirable to try to satisfy *all* these requirements, and in describing the various models which have been proposed, we shall indicate which of the various properties they possess.

The distribution model has been formulated using f_{ij}, a_i, b_j. It is sometimes convenient to introduce dimensionless quantities ρ_{ij}, u_i, v_j defined by

$$\rho_{ij} = v^{-1} f_{ij}, \tag{13}$$

$$u_i = v^{-1} a_i, \tag{14}$$

$$v_j = v^{-1} b_j, \tag{15}$$

so that

$$\sum_i u_i = \sum_j v_j = \sum_{i,j} \rho_{ij} = 1. \tag{16}$$

These normalized quantities are particularly important for the probabilistic interpretation of trip distribution models.

21. Hitchcock Model

The classical transportation problem, formally stated by Hitchcock in his 1941 paper as "The Distribution of a Product from Several Sources to Numerous Localities," can be interpreted as a trip distribution problem in the following way: the traffic desire network is regarded as a bipartite graph in which the m trip origins are numbered $i = 1, 2, ..., m$, and the n trip destinations are numbered $j = 1, 2, ..., n$. If the number of trips originating from i is denoted by a_i, the number of trips with destination j by b_j, and the total number of trips by v, then

$$\sum_1^m a_i = \sum_1^n b_j = v. \tag{1}$$

The cost associated with a single trip from i to j is denoted by c_{ij}, and this is usually determined, on the basis of the first extremal principle discussed in Chap. III, as the cost for the cheapest route between i and j. The number of trips f_{ij} between i and j is then determined, on the basis of the second extremal principle, by minimizing the total network cost C.

The Hitchcock trip distribution model is thus formulated as the linear program

$$f_{ij} \geq 0, \tag{2}$$

$$\sum_j f_{ij} = a_i, \tag{3}$$

$$\sum_i f_{ij} = b_j, \tag{4}$$

$$\sum_i \sum_j c_{ij} f_{ij} = C(\text{min}), \tag{5}$$

with $i = 1, 2, \ldots, m$, and $j = 1, 2, \ldots, n$. The dual of this LP is

$$\alpha_i + \beta_j \leq c_{ij}, \tag{6}$$

$$\alpha_i, \beta_j \quad \text{unrestricted in sign}, \tag{7}$$

$$\sum a_i \alpha_i + \sum b_j \beta_j = V(\text{max}), \tag{8}$$

where the dual variables α_i, β_j are usually called implicit prices. The various algorithms for the solution of this special LP are too well known and documented to bear repetition here, and the interested reader can find excellent and exhaustive discussions in reference [1], Sect. 11, Chap. II or [18], Sect. 17, Chap. III. Instead, we shall briefly interpret the minimum cost solution as an optimal trip distribution, and describe its properties.

An important property of the optimal solution is that, in general, not more than $m + n - 1$ of the mn optimal trip numbers f_{ij}^* can be nonzero. In other words, the optimal trip distribution represents a concentration of the trips on relatively few of the possible O–D desire lines. This concentration could be quite erroneous for certain trip purposes, such as recreation and nonhome-based trips, which are likely to be rather spread out over an area. On the other hand, this concentration is evident for home-based work trips because of the tendency for people to choose to live near where they work. In the afternoon peak period, for example, we might regard the origins of the work-to-home trips comprising, say, 50 zones and the destinations 300 zones. The optimal trip distribution would then be concentrated on about 350 O–D pairs, which represents an average of 7 possible destination zones for each origin zone. It is clear, then, that the chosen subdivision of the area into zones and the trip purpose under consideration will be an important factor in deciding whether the Hitchcock model would be an appropriate distribution model.

The Hitchcock model evidently satisfies the conservation requirements and also automatically ensures nonnegative integral trip numbers.

The model can be forced to satisfy the compressibility constraint by an appropriate but somewhat artificial choice of interzonal costs. Suppose that the destination zones $n-1$ and n are combined. Then compressibility requires that

$$f'_{ij} = f_{ij}, \qquad a'_i = a_i, \qquad b'_j = b_j, \qquad (9)$$

$$f'_{i,n-1} = f_{i,n-1} + f_{in}, \qquad b'_{n-1} = b_{n-1} + b_n, \qquad (10)$$

for $i = 1, 2, ..., m$, and $j = 1, 2, ..., n-2$. To achieve this, the new costs $c'_{i,n-1}$ have to be calculated from the optimal solution of the original problem. If f^*_{ij} denotes the optimal solution of the original primal problem, and α_i^*, β_j^* the optimal solution of the original dual problem, then the new costs are chosen as follows:

$$c'_{ij} = c_{ij}, \qquad (11)$$

$$c'_{i,n-1} = \frac{b_{n-1} c_{i,n-1} + b_n c_{in}}{b_{n-1} + b_n},$$

$$\text{either } f^*_{i,n-1} > 0, \quad f^*_{in} > 0, \quad \text{or } f^*_{i,n-1} = 0, \quad f^*_{in} = 0, \qquad (12)$$

$$c'_{i,n-1} = \frac{b_{n-1}(\alpha_i^* + \beta_{n-1}^*) + b_n(\alpha_i^* + \beta_n^*)}{b_{n-1} + b_n},$$

$$\text{either } f^*_{i,n-1} > 0, \quad f^*_{in} = 0, \quad \text{or } f^*_{i,n-1} = 0, \quad f^*_{in} > 0, \qquad (13)$$

for $i = 1, 2, ..., m$, and $j = 1, 2, ..., n-2$. Equation (12) indicates that if the desire lines between zone i and zones $n-1$ and n are either both used or both not used in the original network, then for the compressed network the cost of a trip between zone i and n is a convex linear combination of the original costs, using the attractions b_{n-1} and b_n as multipliers. It is evident that this combination would not be satisfactory to describe the case when one and only one of the desire lines is used. For suppose $c_{i,n}$ were very large, forcing $f^*_{in} = 0$, but that $c_{i,n-1}$ were sufficiently small so that $f^*_{i,n-1} > 0$. The convex linear combination of $c_{i,n-1}$ and $c_{i,n}$ could well be large enough to incorrectly force $f'^*_{i,n-1} = 0$. In this particular case, (13) becomes

$$c'_{i,n-1} = \frac{b_{n-1} c_{i,n-1} + b_n(\alpha_i^* + \beta_n^*)}{b_{n-1} + b_n}, \qquad (14)$$

because of the complementary slackness relations.

That this choice of new costs gives the required compressibility of the model is most easily verified by appeal to duality theory. It is claimed that if $f_{ij}^*, \alpha_i^*, \beta_j^*$ are the optimal values of the primal and dual variables for the original network, then

$$f_{ij}'^* = f_{ij}^*, \tag{15}$$

$$f_{i,n-1}'^* = f_{i,n-1}^* + f_{in}^*, \tag{16}$$

$$\alpha_i'^* = \alpha_i^*, \qquad \beta_j'^* = \beta_j^*, \tag{17}$$

$$\beta_{n-1}'^* = \frac{b_{n-1}\beta_{n-1}^* + b_n \beta_n^*}{b_{n-1} + b_n}, \tag{18}$$

for $i = 1, 2, \ldots, m$, and $j = 1, 2, \ldots, n-2$, give the optimal values of the primal and dual variables for the compressed network. It is not difficult to check that these values give primal and dual feasible solutions and that the complementary slackness relations

$$f_{i,n-1}'^* > 0 \Rightarrow \alpha_i'^* + \beta_{n-1}'^* = c_{i,n-1}^*, \tag{19}$$

and

$$\alpha_i'^* + \beta_{n-1}'^* < c_{i,n-1}' \Rightarrow f_{i,n-1}'^* = 0, \tag{20}$$

are valid, giving optimality. It is an immediate consequence of (11) that since $C^* = V^*$ and $C'^* = V'^*$,

$$C'^* = V'^* = \sum_1^m a_i' \alpha_i'^* + \sum_1^{n-1} b_j' \beta_j'^*,$$

$$= \sum_1^m a_i \alpha_i^* + \sum_1^n b_j \beta_j^*,$$

$$= V^* = C^*,$$

i.e., the optimal network cost is unchanged.

Thus, with appropriate choice of interzonal costs, the Hitchcock model can be made to satisfy the compressibility requirements. However, it does not possess the separability property.

22. Entropy Models

(a) Network Entropy

The concept of entropy, familiar in thermodynamics and information theory, provides a useful unifying basis for a class of distribution models, of which the gravity model has been one of the most commonly used in transportation planning. Murchland [2] seems to have been the first to show explicitly that the gravity model can be formulated as the solution of an equivalent maximization problem and subsequently many authors have formulated various entropy maximizing models [3].

It is convenient to introduce the concept of network entropy in terms of the dimensionless quantities ρ_{ij} obtained by dividing f_{ij} by v as in (13), Sect. 20. The quantity ρ_{ij} is interpreted as the joint probability of a trip being produced at zone i and attracted to zone j, implying the constraints

$$0 \leqslant \rho_{ij} \leqslant 1, \tag{1}$$

$$\sum_{i,j} \rho_{ij} = 1. \tag{2}$$

The *network entropy* is then defined as

$$H = -\sum_{i,j} \rho_{ij} \ln \rho_{ij}. \tag{3}$$

This definition is the same as that used in thermodynamics and information theory and the familiar conventions will be adopted here, e.g., $\rho_{ij} \ln \rho_{ij}$ is taken as zero when $\rho_{ij} = 0$. A single term in the summation in (3) has the following properties:

$$-\rho_{ij} \ln \rho_{ij} = 0, \quad \text{for} \quad \rho_{ij} = 0, 1, \tag{4}$$

$$-\rho_{ij} \ln \rho_{ij} > 0, \quad \text{for} \quad 0 < \rho_{ij} < 1, \tag{5}$$

$$-\rho_{ij} \ln \rho_{ij} \quad \text{has a maximum of} \quad e^{-1} \quad \text{when} \quad \rho_{ij} = e^{-1}, \tag{6}$$

$$-\rho_{ij} \ln \rho_{ij} \quad \text{is concave,} \quad \text{for} \quad 0 < \rho_{ij} < 1. \tag{7}$$

Properties (6) and (7) follow from

$$\frac{d}{d\rho_{ij}} (-\rho_{ij} \ln \rho_{ij}) = -\ln \rho_{ij} - 1, \tag{8}$$

and

$$\frac{d^2}{d\rho_{ij}^2} (-\rho_{ij} \ln \rho_{ij}) = -\frac{1}{\rho_{ij}}. \tag{9}$$

The network entropy is a measure of uncertainty, and entropy distribution models are based on the principle that *an equilibrium distribution maximizes the entropy*. The significance of this principle will be illustrated by a simple example which hardly merits description as a distribution model.

Suppose there is *no* knowledge about the distribution of trips for a network with m origin and n destination nodes. Then the entropy model is obtained by maximizing the network entropy (3) subject to (2). It is instructive to do this by using a Lagrange multiplier v and defining a Lagrangian function

$$\mathscr{L}(\rho_{ij}, v) = -\sum \rho_{ij} \ln \rho_{ij} - v(\sum \rho_{ij} - 1). \tag{10}$$

Extreme values of this function correspond to

$$\frac{\partial}{\partial \rho_{ij}} \mathscr{L} = -\ln \rho_{ij} - 1 - v = 0, \tag{11}$$

and

$$\frac{\partial}{\partial v} \mathscr{L} = -\sum \rho_{ij} + 1 = 0, \tag{12}$$

which, when solved for the equilibrium values ρ_{ij}^*, give the unique solutions

$$\rho_{ij}^* = 1/mn \tag{13}$$

$$H^* = H_{\max} = \ln mn. \tag{14}$$

It is essential to check that the equilibrium values ρ_{ij}^* given by (13) satisfy the nonnegativity restrictions (1).

This example illustrates the technique used in maximizing network entropy; the result is obvious. In the absence of any knowledge, maximizing network entropy distributes the trips evenly through the network, the expected number of interzonal and intrazonal trips being all equal and independent of the zones.

(b) Proportional Model

One of the simplest distribution models is the proportional model which is obtained by maximizing the network entropy $H = -\sum \rho_{ij} \ln \rho_{ij}$ subject to the restrictions

$$\sum_{j=1}^{n} \rho_{ij} = u_i, \qquad \sum_{i=1}^{m} \rho_{ij} = v_j, \tag{15}$$

in addition to the constraints

$$0 \leqslant \rho_{ij} \leqslant 1, \qquad \sum_{i,j} \rho_{ij} = 1. \tag{16}$$

The quantities u_i, v_j are the nonnegative normalized trip productions and attractions defined by (14) and (15), Sect. 20. The model ignores any interzonal trip costs.

(i) Lagrange Multiplier Derivation

To obtain the equations defining the model, m Lagrange multipliers, λ_i, and n multipliers, μ_j, are introduced, as well as the multiplier v, to define the Lagrangian function

$$\mathcal{L}(\rho_{ij}, \lambda_i, \mu_j, v) = -\sum_{i,j} \rho_{ij} \ln \rho_{ij} - \sum_i \lambda_i \left(\sum_j \rho_{ij} - u_i \right)$$
$$- \sum_j \mu_j \left(\sum_i \rho_{ij} - v_j \right) - v \left(\sum_{i,j} \rho_{ij} - 1 \right). \quad (17)$$

Extreme values correspond to

$$\frac{\partial}{\partial \rho_{ij}} \mathcal{L} = -\ln \rho_{ij} - 1 - \lambda_i - \mu_j - v = 0, \qquad (18)$$

$$\frac{\partial}{\partial \lambda_i} \mathcal{L} = -\sum_j \rho_{ij} + u_i = 0, \qquad (19)$$

$$\frac{\partial}{\partial \mu_j} \mathcal{L} = -\sum_i \rho_{ij} + v_j = 0, \qquad (20)$$

$$\frac{\partial}{\partial v} \mathcal{L} = -\sum_{i,j} \rho_{ij} + 1 = 0. \qquad (21)$$

The solution to (18) is

$$\rho_{ij}^* = \exp(-1 - v - \lambda_i - \mu_j), \qquad (22)$$

with the remaining equations forcing

$$u_i = \exp(-1 - v - \lambda_i) \sum_j \exp(-\mu_j), \qquad (23)$$

$$v_j = \exp(-1 - v - \mu_j) \sum_i \exp(-\lambda_i), \qquad (24)$$

$$1 = \exp(-1 - v) \sum_{i,j} \exp(-\lambda_i - \mu_j). \qquad (25)$$

The multipliers are eliminated by taking the product of (23) and (24) and dividing by (25) to give

$$u_i v_j = \exp(-1 - v - \lambda_i - \mu_j) = \rho_{ij}^*. \qquad (26)$$

In terms of the unnormalized quantities (and dropping the asterisk) we have

$$f_{ij} = a_i b_j / v. \tag{27}$$

The proportional model is characterized as an even distribution of trips subject to given productions and attractions. In the absence of any knowledge of the geography of the network—e.g., of interzonal distances—the proportional model maximizes the entropy giving an equilibrium distribution in which the trips between zones are distributed strictly in accord with the proportion of the trips produced at and attracted to the zones. The joint probability of a trip being produced at zone i and attracted to zone j is expressed as the product of u_i, the probability of a trip being produced at i, and v_j, the probability of a trip being attracted to j.

(ii) A PROBABILISTIC INTERPRETATION

We now rederive (26) using a purely probabilistic argument which gives further insight into the entropy concept. Let x_i denote the probability that origin i is chosen by a single traveler and y_j the probability that destination j is chosen. Assume that the choice of origin and destination are made independently of one another. The joint probability of choosing origin i and destination j and hence the trip (i, j) is therefore $x_i y_j$. When a_i, b_j, and v are all integers, the joint probability that a_1 travelers independently pick origin $1, \ldots, a_i$ travelers independently pick origin $i, \ldots, b_j$ travelers independently pick destination $j, \ldots,$ and so on, is

$$x_1^{a_1} x_2^{a_2} \cdots x_i^{a_i} \cdots x_m^{a_m} y_1^{b_1} \cdots y_j^{b_j} \cdots y_n^{b_n}$$

$$= [x_1^{u_1} x_2^{u_2} \cdots x_m^{u_m} y_1^{v_1} \cdots y_n^{v_n}]^v, \tag{28}$$

since $a_i = v u_i$ and $b_j = v v_j$. In other words, this joint probability is the vth power of a probability

$$P(x, y) = x_1^{u_1} x_2^{u_2} \cdots x_m^{u_m} y_1^{v_1} y_2^{v_2} \cdots y_n^{v_n}, \tag{29}$$

which is another way of saying that the joint probability in (28) is the product of v terms, i.e., the probability of the simultaneous realization of v independent events whose probability of occurrence is given by (29).

Notice that the probability of indistinguishable arrangements which are obtained by interchanging travelers between different nodes and adding terms of the form of (28) as many times as there are inter-changes which lead to the same sequences $(a_1, a_2, ..., a_m)(b_1, b_2, ..., b_n)$ is *not* the probability that we discuss.

Of course, some trip from i or to j must be chosen by each traveler so that $\sum_i x_1 = 1$, $\sum_j y_j = 1$. This implies that the sum of joint probabilities of an (i, j) trip must equal one, or

$$\sum_{i,j} x_i y_j = 1. \tag{30}$$

We will only be interested in strictly positive values of x_i and y_j that satisfy (30); otherwise, (28) and (29) are zero.

We shall now show that if there exist x_i and y_j that satisfy the probability constraint of (30) and, furthermore, if there exists ρ_{ij} that satisfy the conservation laws, (15), then we shall always have

$$-H(\rho) = \sum_{i,j} \rho_{ij} \ln \rho_{ij} \geqslant \sum_i u_i \ln x_i + \sum_j v_j \ln y_j = \ln P(x, y), \tag{31}$$

or,

$$P(x, y) \leqslant \exp[-H(\rho)],$$

for feasible x_i, y_j, and ρ_{ij}. Furthermore, the equality is attainable in the sense that for feasible, x_i, y_j, and ρ_{ij}, we have

$$\max_\rho H(\rho) = \min_{x,y} \ln\left(\frac{1}{P(x, y)}\right), \tag{32}$$

where it is understood that x, y, and ρ are constrained as before.

In order to obtain this result, we make use of a well-known inequality relating arithmetic and geometric means: for arbitrary positive numbers α_k which satisfy the normality condition

$$\alpha_1 + \alpha_2 + \cdots + \alpha_p = 1, \tag{33}$$

and for arbitrary nonnegative numbers $z_1, z_2, ..., z_p$, we have

$$\alpha_1 z_1 + \alpha_2 z_2 + \cdots + \alpha_p z_p \geqslant z_1^{\alpha_1} z_2^{\alpha_2} \cdots z_p^{\alpha_p}, \tag{34}$$

with equality holding if and only if $z_1 = z_2 = \cdots = z_p$. This geometric inequality can be written in a slightly more useful form for our purposes, by taking the logarithm of both sides of (34). This yields the inequality

$$\ln\left[\sum_k \alpha_k z_k\right] \geqslant \sum_k \alpha_k \ln z_k, \tag{35}$$

since we know that $\ln a \geqslant \ln b$, if $a \geqslant b > 0$. By writing this expression in terms of new variables

$$t_k = \ln \alpha_k z_k = \ln \alpha_k + \ln z_k, \tag{36}$$

we now obtain

$$\ln \left[\sum_k \exp(t_k) \right] \geqslant \sum_k \alpha_k (t_k - \ln \alpha_k), \tag{37}$$

or, alternatively,

$$-\sum_k \alpha_k \ln \alpha_k \leqslant -\sum_k \alpha_k t_k + \ln \left[\sum_k \exp(t_k) \right], \tag{38}$$

for arbitrary real t_k or arbitrary nonnegative z_k.

We are now in a position to show that the inequality (31) holds for feasible x, y, ρ. The right-hand side of (31) is

$$\ln P(x, y) = \left[\sum_i u_i \ln x_i + \sum_j v_j \ln y_j \right]. \tag{39}$$

Since we assume that the ρ_{ij} on the left-hand side of (31) are feasible, we can substitute the left-hand sides of (15) for u_i and v_j in (39). This yields the expression

$$\ln P(x, y) = \left[\sum_i \sum_j \rho_{ij} \ln x_i + \sum_j \sum_i \rho_{ij} \ln y_j \right]. \tag{40}$$

Since each ρ_{ij} premultiplies both a $\ln x_i$ and a $\ln y_j$ term, we can collect $x_i y_j$ products to obtain

$$\ln P(x, y) = \sum_{i,j} \rho_{ij} \ln(x_i y_j) = \sum_{i,j} \rho_{ij} z_{ij}, \tag{41}$$

where

$$z_{ij} = \ln(x_i y_j) < 0$$

defines z_{ij} for all nonnegative x_i, y_j. We now make use of (38) by substituting z_{ij} for the typical t_k term, substituting ρ_{ij} for the typical α_k term, and using a double rather than a single summation. Since the logarithm of (30) is

$$\ln \left[\sum_{i,j} x_i y_j \right] = \ln \left[\sum_{i,j} \exp(z_{ij}) \right] = \ln(1) = 0, \tag{42}$$

we can write the right-hand side of (41) in the form

$$-\ln P(x, y) = -\sum_{i,j} \rho_{ij} z_{ij} + \ln \left[\sum_{i,j} \exp(z_{ij}) \right], \tag{43}$$

and make use of the inequality of (38),

$$-\sum_{i,j} \rho_{ij} z_{ij} + \ln\left[\sum_{i,j} \exp(z_{ij})\right] \geqslant -\sum \rho_{ij} \ln \rho_{ij}, \qquad (44)$$

to obtain the desired inequality

$$-\ln P(x,y) \geqslant H(\rho). \qquad (45)$$

It is now trivial to show that the maximum and minimum values in (32) are actually attained when

$$x_i^* = v^{-1} a_i, \qquad y_j^* = v^{-1} b_j, \qquad \rho_{ij}^* = u_i v_j. \qquad (46)$$

Notice, first of all, that these values are feasible for the probability constraint and the flow conservation equations. Hence, we know that the inequality of (31) must hold. But the entropy is

$$H(\rho^*) = -\sum_{i,j} u_i v_j \ln(u_i v_j), \qquad (47)$$

whereas

$$-\ln P(u,v) = -\sum_i u_i \ln u_i - \sum_j v_j \ln v_j$$

$$= -\sum_{i,j} u_i v_j (\ln u_i + \ln v_j)$$

$$= -\sum_{i,j} u_i v_j \ln(u_i v_j) = H(\rho^*), \qquad (48)$$

and the equality has been demonstrated.

(iii) MODEL PROPERTIES

It is evident that the proportional model satisfies the conservation requirements, and that the trip numbers are nonnegative, though not necessarily integral. In a practical application, round-off to integers would be satisfactory. The proportional model also has the compressibility property as is easily verified by comparing (4)–(7), Sect. 20, with the following:

$$a_i' b_j'/v' = a_i b_j/v, \qquad \text{with} \quad v' = v, \qquad (49)$$

$$a_{n-1}' b_j'/v' = a_{n-1} b_j/v + a_n b_j/v, \qquad (50)$$

$$a_i' b_{n-1}'/v' = a_i b_{n-1}/v + a_i b_n/v, \qquad (51)$$

$$a_{n-1}' b_{n-1}'/v' = a_{n-1} b_{n-1}/v + a_{n-1} b_n/v + a_n b_{n-1}/v + a_n b_n/v, \quad (52)$$

for $i,j = 1, 2, \ldots, n-2$.

The model is also separable. Again this can be checked by comparing (12), Sect. 20, with the following:

$$
\begin{aligned}
a_i' b_j'/v' &= \frac{(a_i - f_{in})(b_j - f_{nj})}{v - a_n - b_n + f_{nn}} \\
&= \frac{(a_i - a_i b_n/v)(b_j - a_n b_j/v)}{v - a_n - b_n + a_n b_n/v} \\
&= \frac{a_i(v - b_n) b_j(v - a_n)}{v(v - a_n)(v - b_n)} \\
&= a_i b_j/v,
\end{aligned}
\tag{53}
$$

for $i, j = 1, 2, \ldots, n-1$.

Consider as a numerical example a four-centroid traffic desire network with the following data:

Centroid	i	1	2	3	4
Production	a_i	25	25	15	35
Attraction	b_i	40	20	30	10.

The total number of trips is $\sum a_i = \sum b_i = 100$, so that ρ_{ij} will be equal to percentages. The trip distribution matrix obtained from the proportional model is, by (27),

$$
[f_{ij}] =
\begin{array}{c|cccc|c}
 & 1 & 2 & 3 & 4 & a_i \\
\hline
1 & 10 & 5 & 7.5 & 2.5 & 25 \\
2 & 10 & 5 & 7.5 & 2.5 & 25 \\
3 & 6 & 3 & 4.5 & 1.5 & 15 \\
4 & 14 & 7 & 10.5 & 3.5 & 35 \\
\hline
b_j & 40 & 20 & 30 & 10 & 100 = v.
\end{array}
\tag{54}
$$

To verify compressibility, suppose that zones 3 and 4 are combined. The third and fourth rows and columns of this matrix are added together to give the new distribution matrix

$$
[f_{ij}'] =
\begin{array}{c|ccc|c}
 & 1 & 2 & 3 & a_i' \\
\hline
1 & 10 & 5 & 10 & 25 \\
2 & 10 & 5 & 10 & 25 \\
3 & 20 & 10 & 20 & 50 \\
\hline
b_j' & 40 & 20 & 40 & 100 = v'.
\end{array}
\tag{55}
$$

It can be readily verified that for this compressed matrix,

$$f'_{ij} = a_i' b_j'/v', \tag{56}$$

as given by the proportional model.

To verify separability, suppose that zone 4 is removed giving the distribution matrix

$$
[f'_{ij}] = \begin{array}{c} \\ 1 \\ 2 \\ 3 \end{array}
\begin{array}{ccc}
1 & 2 & 3 \\
\left[\begin{array}{ccc} 10 & 5 & 7.5 \\ 10 & 5 & 7.5 \\ 6 & 3 & 4.5 \end{array}\right]
\end{array}
\begin{array}{c} a_i' \\ 22.5 \\ 22.5 \\ 13.5 \end{array}
\tag{57}
$$

$$b_j' \quad 26 \quad 13 \quad 19.5 \qquad 58.5 = v'.$$

Again it can be checked that

$$f'_{ij} = a_i' b_j'/v', \tag{58}$$

as given by the proportional model.

The proportional model is a simple distribution model which possesses all the desired properties (except integrality of trip numbers). It is of little practical importance for distributing trips, because it takes no account of the geographical disposition of the zones of the study area.

(c) Mean Trip Length Model

Travel patterns indicate a general preference for shorter trips than longer trips and the *trip length frequency distributions* obtained from travel inventories have proved a reliable characterization of travel in a particular study area. The frequency distributions differ for differing purposes, longer trips being more common, for example, for recreation trips than for home-based work trips. To add more realism to the proportional model, we include just one parameter of the trip length frequency distribution, namely the *mean trip length*, and we fix this as an additional constraint on our distribution. As has been noticed previously, trip "length" is sometimes measured in terms of distance, sometimes as travel time, and we continue our practice of using the more general term cost. If c_{ij} is the given fixed cost of a trip interchange produced at node i and attracted to node j, the total network cost is

$$C = \sum_{i,j} f_{ij} c_{ij} \tag{59}$$

and the mean cost of a trip can be defined as

$$c = \sum_{i,j} \rho_{ij} c_{ij}. \tag{60}$$

We now impose this relation, with c regarded as fixed, as a constraint on our distribution model in addition to the given productions and attractions. In detail, this entropy model can be formulated as:

$$\text{maximize } H = -\sum_{i,j} \rho_{ij} \ln \rho_{ij}, \tag{61}$$

subject to

$$\sum_{j} \rho_{ij} = u_i, \tag{62}$$

$$\sum_{i} \rho_{ij} = v_j, \tag{63}$$

$$\sum_{i,j} \rho_{ij} = 1, \tag{64}$$

$$\sum_{i,j} \rho_{ij} c_{ij} = c. \tag{65}$$

Such a model has been considered by Sasaki [4] and Tomlin and Tomlin [5].

In the usual way, a Lagrangian is first defined as

$$\mathscr{L}(\rho_{ij}, \lambda_i, \mu_j, \nu, \gamma)$$
$$= -\sum_{i,j} \rho_{ij} \ln \rho_{ij} - \sum_{i} \lambda_i \left(\sum_{j} \rho_{ij} - u_i \right) - \sum_{j} \mu_j \left(\sum_{i} \rho_{ij} - v_j \right)$$
$$- \nu \left(\sum_{i,j} \rho_{ij} - 1 \right) - \gamma \left(\sum_{i,j} \rho_{ij} c_{ij} - c \right), \tag{66}$$

which is the same as (17) except for the additional term with multiplier γ. The optimal values ρ_{ij}^* are given by

$$\rho_{ij}^* = \exp(-1 - \nu - \lambda_i - \mu_j - \gamma c_{ij}), \tag{67}$$

where

$$u_i = \exp(-1 - \nu - \lambda_i) \sum_{j} \exp(-\mu_j - \gamma c_{ij}), \tag{68}$$

$$v_j = \exp(-1 - \nu - \mu_j) \sum_{i} \exp(-\lambda_i - \gamma c_{ij}), \tag{69}$$

$$1 = \exp(-1 - \nu) \sum_{i,j} \exp(-\lambda_i - \mu_j - \gamma c_{ij}), \tag{70}$$

$$c = \exp(-1 - \nu) \sum_{i,j} c_{ij} \exp(-\lambda_i - \mu_j - \gamma c_{ij}). \tag{71}$$

It is evident from (67) and (68) that $0 < \rho_{ij}^* < 1$, provided the multipliers are finite.

It is not possible to eliminate the multipliers and obtain an explicit expression for ρ_{ij}^*; instead an iterative or model calibration procedure is used. Equation (67) has the form

$$\rho_{ij} = x_i w_{ij} y_j, \tag{72}$$

where

$$w_{ij} = \exp(-\gamma c_{ij}), \tag{73}$$

and the asterisk has been dropped. The x_i, y_j, γ have to be determined from the constraints

$$\sum_j x_i w_{ij} y_j = u_i, \tag{74}$$

$$\sum_i x_i w_{ij} y_j = v_j, \tag{75}$$

$$\sum_{i,j} c_{ij} x_i w_{ij} y_j = c. \tag{76}$$

The iterative procedure begins with a trial value of γ from which w_{ij} is determined from (73). A convenient starting point is to take the initial value of γ as c^{-1}. Then successive values of y_j and x_i are chosen in order to satisfy (74) and (75). Starting values $y_j^{(1)}$ are first chosen (say $y_j^{(1)} = v_j$) and $x_i^{(1)}$ determined from

$$x_i^{(1)} = u_i \left[\sum_j w_{ij} y_j^{(1)} \right]^{-1}, \tag{77}$$

in order to satisfy (74). Next $y_j^{(2)}$ is determined from

$$y_j^{(2)} = v_j \left[\sum_i x_i^{(1)} w_{ij} \right]^{-1}, \tag{78}$$

and, in general,

$$y_j^{(k)} = v_j \left[\sum_i x_i^{(k-1)} w_{ij} \right]^{-1}, \tag{79}$$

$$x_i^{(k)} = u_i \left[\sum_i w_{ij} y_j^{(k)} \right]^{-1}, \tag{80}$$

giving the kth iterate

$$\rho_{ij}^{(k)} = x_i^{(k)} w_{ij} y_j^{(k)}. \tag{81}$$

Termination of this procedure can be programmed in several ways. One can require, for example, that for some small number δ, the iteration ends when

$$\max_{(i,j)} [|\rho_{ij}^{(k)} - \rho_{ij}^{(k-1)}|] \leqslant \delta. \tag{82}$$

When this iteration has been completed, (76) has to be checked and a new value of γ chosen in order to reduce any discrepancy. With this new value, the iteration process is carried through again. The calibration of the model to fit the given mean trip length is a rather trial-and-error process and rapid conversion to a satisfactory model requires considerable judgement in gauging the effect of the given distribution of interzonal costs.

The justification for the iterative procedure in determining the x_i and y_j is found in a theorem by Sinkhorn [6] which states that the procedure converges to a unique solution. An important requirement is that the values of w_{ij} must be positive, which is certainly satisfied in this model. The theorem statement also allows for the obvious fact that the solution for x_i and y_j is unique up to an arbitrary factor; multiplying x_i by a constant and dividing y_j by this constant obviously leaves the result unchanged.

As a distribution model, the mean trip length model satisfies the conservation equations and its elements are positive (although not necessarily integral). References [4] and [5] contain interesting applications of the model to actual data.

(d) Gravity Model

The gravity model has been the most widely used distribution model in transportation planning and has formed the basis of traffic predictions for many cities. It is not the purpose here to describe the model in great detail but rather to emphasize its relation to the network entropy concept.

The gravity model can be regarded as a logical extension of the mean trip length model. Instead of fixing just one parameter—the mean—of the trip length frequency distribution, the complete distribution is fitted. For simplicity we shall let $\rho(t)$ be the percentage of trips with trip lengths c_{ij} in the range $t \leqslant c_{ij} < t + \Delta t$, where Δt is a suitably chosen increment. For example, c_{ij} may be measured in minutes and the trip length frequency distribution represented by a histogram with $\Delta t = 2$ minutes. We write

$$\rho(t) = \sum{}' \rho_{ij}, \tag{83}$$

where the summation $\sum'$ is over all ρ_{ij} for which $t \leqslant c_{ij} < t + \Delta t$. For each value of t a Lagrange multiplier $\gamma(t)$ is defined so that, in generalization of (66), the Lagrangian function becomes

$$\mathscr{L}(\rho_{ij}, \lambda_i, \mu_j, \nu, \gamma(t))$$

$$= -\sum_{i,j} \rho_{ij} \ln \rho_{ij} - \sum_i \lambda_i \left(\sum_j \rho_{ij} - u_i \right)$$

$$- \sum_j \mu_j \left(\sum_i \rho_{ij} - v_j \right) - \nu \left(\sum_{i,j} \rho_{ij} - 1 \right) - \sum_t \gamma(t) \left(\sum' \rho_{ij} - \rho(t) \right).$$

$$(84)$$

As in (67)–(76), the maximum entropy principle leads to equilibrium values of

$$\rho_{ij} = x_i w_{ij} y_j, \qquad (85)$$

where

$$w_{ij} = \exp(-\gamma(t)), \qquad (86)$$

and x_i, y_j, and $\gamma(t)$ are determined from

$$\sum_j x_i w_{ij} y_j = u_i, \qquad (87)$$

$$\sum_i x_i w_{ij} y_j = v_j, \qquad (88)$$

$$\sum' x_i w_{ij} y_j = \rho(t). \qquad (89)$$

The function $\exp(-\gamma(t))$ is sometimes called the deterrence function and may be chosen as a simple form with parameters determined from (89) or given by a table of values. By analogy with Newton's law, the deterrence function was originally chosen as $1/t^2$, corresponding to $\gamma(t) = 2 \ln t$. However, improved fits were obtained by taking $\gamma(t) = at + b \ln t$ with a and b as parameters to be determined by calibration. With this sort of generalization, the model no longer resembles Newton's law of gravity, and the distribution model is often called an interactance model.

In Table 22.1 we illustrate possible forms of the deterrence function. The data in the second column are measured values for home-based work trips for Washington, D.C. [12], with t the travel time in minutes. The third and fourth columns of the table illustrate the fit one can get with $\gamma(t) = 2 \ln t$ and $\gamma(t) = 0.14t$, respectively. Suitable multiplicative factors have been used for comparison with the measured values, which are only determined to within an arbitrary factor.

TABLE 22.1   DETERRENCE FUNCTION FOR GRAVITY MODEL

t (min)	Deterrence function	$5000 \times t^{-2}$	$2000 \times$ $\exp(-0.14t)$
5	1000	2000	1000
10	500	500	500
15	205	220	240
20	120	120	120
25	82	80	60
30	53	55	30
35	34	41	15
40	19	31	7
45	9	25	4
50	3	20	2
55	1	16	1

In practice, the calibration of the gravity model is an intricate procedure which, especially for large networks, becomes more of an art than a science. Balancing the row and column sums to satisfy (87) and (88) is achieved by a matrix scaling method similar to that already described. The deterrence function is represented by travel-time or friction factors, called F-factors, and these are modified by special zone-to-zone adjustment factors, called K-factors, until a satisfactory fit with the trip length frequency distribution is obtained. The procedure is rather heuristic, and the model structure tends to become obscured. A difficulty which may arise from zero or small elements in the trip table is the oscillation and nonconvergence of the scaling procedure. This phenomenon is now understood as pointed out in the discussion to reference [6].

It is interesting to transform the notation used here to the familiar formulation of the gravity model. For a particular trip purpose, the gravity model is usually defined by

$$f_{ij}^{(k)} = a_i F_{ij} K_{ij} b_j^{(k)} \left[\sum_j F_{ij} K_{ij} b_j^{(k)} \right]^{-1}, \qquad (90)$$

$$b_j^{(1)} = b_j, \qquad (91)$$

$$b_j^{(k)} = b_j b_j^{(k-1)} \left[\sum_i f_{ij}^{(k-1)} \right]^{-1}, \qquad (92)$$

where

$f_{ij}^{(k)}$ = elements of the trip table at the kth iteration,

F_{ij} = travel-time factors,

K_{ij} = adjustment factors.

If we now define

$$f_{ij}^{(k)} = v\rho_{ij}^{(k)}, \tag{93}$$

$$F_{ij}K_{ij} = vw_{ij}, \tag{94}$$

$$b_j^{(k)} = y_j^{(k)}, \tag{95}$$

$$a_i\left[\sum_j F_{ij}K_{ij}b_j^{(k)}\right]^{-1} = x_i^{(k)}, \tag{96}$$

then

$$\rho_{ij}^{(k)} = x_i^{(k)} w_{ij} y_j^{(k)}, \tag{97}$$

with

$$x_i^{(k)} = u_i\left[\sum_j w_{ij} y_j^{(k)}\right]^{-1}, \tag{98}$$

and from (94)

$$y_j^{(k)} = v_j y_j^{(k-1)}\left[\sum_j x_i^{(k-1)} w_{ij} y_j^{(k-1)}\right]^{-1}$$

$$= v_j\left[\sum_j x_i^{(k-1)} w_{ij}\right]^{-1}. \tag{99}$$

These are precisely (79)–(81).

Like the mean trip length model, the gravity model satisfies the conservation equations, has nonnegative (but not necessarily integral) elements, but does not possess the compressibility or separability properties. Despite its almost universal use, the model is now regarded by some with suspicion [8].

23. Opportunity Models

Opportunity trip distribution models come second only to the gravity model in popularity among transportation planners. Stouffer [9] first introduced the concept of opportunities as a useful variable in describing travel behavior and applied the concept to the residential mobility of people.

For a traffic desire network, the opportunities are determined as possible destinations. For a particular origin zone, the intervening opportunities are calculated from an ordering of all destination centroids according to their costs (distances, travel times etc.) from the origin. If, as in our notation, we define b_j as the number of trip destinations at centroid j, then the number of intervening opportunities between (and including) origin i and destination j is given by

$$B_{ij} = \sum{}' b_l, \tag{1}$$

where $\sum'$ is the sum over all centroids l for which $c_{il} \leqslant c_{ij}$.

We shall describe two trip distribution models using the concept of opportunities.

(a) Intervening Opportunities Model

The intervening opportunities trip distribution model was developed for the Chicago Area Transportation Study (CATS). The model is based on the following assumption: the conditional probability that a trip having originated at zone i has a destination beyond j is given by $\exp(-L_i B_{ij})$, where B_{ij} is the number of intervening opportunities between i and j, and L_i is a constant. This exponential distribution is familiar in the kinetic theory of gases (when the probability refers to a molecule traveling a distance without colliding), and the theory of radioactive decay (when the probability refers to an atom not decaying in a certain time).

The quantity $1/L_i$, analogous to the mean free path between molecular collisions or mean life of a radioactive atom, is interpreted as the mean number of intervening opportunities not accepted. The estimated trip numbers from i beyond j are given by the formula

$$\sum_{c_{il} > c_{ij}} f_{il} = a_i \exp(-L_i B_{ij}). \tag{2}$$

The actual trip numbers between i and j are obtained by differencing, or more explicitly, as

$$f_{ij} = a_i \left[\exp(-L_i \sum{}'' b_l) - \exp(-L_i \sum{}' b_l) \right], \tag{3}$$

where $\sum''$ is the sum over all centroids l for which $c_{il} < c_{ij}$, i.e., without the equality sign used in determining $\sum'$.

This is the mathematical statement of the intervening opportunities trip distribution model. It is evident that the f_{ij} are positive (although not necessarily integral), but that the model does not satisfy the conservation equations

$$\sum_j f_{ij} = a_i, \tag{4}$$

$$\sum_i f_{ij} = b_j. \tag{5}$$

In practice, (4) will be approximately true, and the row balances implied by (5) are achieved by an iterative scheme

$$f_{ij}^{(k)} = a_i \left[\exp\left(-L_i \sum_l'' b_l^{(k)} \right) - \exp\left(-L_i \sum_l' b_l^{(k)} \right) \right], \tag{6}$$

$$b_j^{(1)} = b_j, \tag{7}$$

$$b_j^{(k)} = b_j b_j^{(k-1)} \left[\sum_i f_{ij}^{(k-1)} \right]^{-1}, \tag{8}$$

similar to that used for the gravity model (see (90)–(92), Sect. 22, and also Problem 7, Sect. 27). The final calibration of the model requires the choice of appropriate L-factors L_i which may differ from zone to zone and vary for different purposes. In fact, there is evidence from opportunity curves drawn from data for various cities that the linear term in B_{ij} appearing in the exponent in (2) is inadequate, and the curves are better represented by a cubic.

From the complexity of the intervening opportunities model, it is evident that as a trip distribution model it does not possess the compressibility or separability properties.

(b) Preferencing Model

An opportunity trip distribution model which satisfies the conservation equations without any iterative balancing has been developed by Kirby [10] using the preference concept familiar in utility theory. Each origin zone is considered to rate the available trip destinations in order of preference, according to interzonal costs. The problem of determining an optimal trip distribution for given preferences is then analogous to the marriage problem [11] in which each man of a group rates a group of women in order of preference, and an optimal pairing of couples is sought.

To describe the preferencing model, we first consider our network as a "blown up" bipartite network with v origin nodes and v destination nodes with one trip originating from each origin zone. The distribution of trips then corresponds to a pairing (or marrying off!) of the origin and destination zones.

Each origin i rates the v possible destinations j in order of preference, and likewise each destination j orders the possible origins. A trip distribution is a 1–1 mapping s of the origins onto the destinations so that $s(i)$ is the destination of the trip originating from i. A stable trip distribution has the following property: if origin i prefers destination $s(h)$ to $s(i)$, then destination $s(h)$ prefers origin h to origin i. The distribution would be unstable if i prefers $s(h)$ to $s(i)$, and $s(h)$ prefers i to h, for then a swap of destinations would be mutually preferable. In general, there are many stable distributions, and among these we seek distributions which are optimal. It is necessary to distinguish between origin optimal and destination optimal. An origin-optimal distribution is a stable distribution for which each origin attains its highest preference destination consistent with stability. More precisely, s is said to be *origin optimal* if, for any stable distribution r, each origin i prefers $s(i)$ (or is possibly indifferent) to $r(i)$. Likewise for a *destination-optimal* distribution, each destination attains its highest preference origin consistent with stability. Interesting results in preference theory imply that there always exists just one origin-optimal distribution and just one destination-optimal distribution, and if a distribution is both origin optimal and destination optimal, it is the only stable distribution.

The following example will illustrate these ideas. Suppose that there are $v = 7$ trips to be distributed between 7 origins and 7 destinations, and suppose that the origins rate the destinations in the orders of preference in Table 23.1. The elements of this table are interpreted as follows: origin 5 rates destination 7 as first preference, destination 6 as

TABLE 23.1 Origin Preference

Origins	Destinations						
	1	2	3	4	5	6	7
1	5	3	4	2	1	7	6
2	1	3	2	6	7	4	5
3	5	6	7	3	4	2	1
4	7	5	6	1	2	4	3
5	5	3	4	6	7	2	1
6	2	3	1	6	7	5	4
7	6	5	7	4	3	1	2

second preference, destination 2 as third preference, and so on. Likewise, suppose that the destinations rate the origins in the orders of preference given in Table 23.2. The algorithm for determining the origin-optimal

TABLE 23.2   DESTINATION PREFERENCE

Desti-nations	Origins						
	1	2	3	4	5	6	7
1	7	5	6	3	4	2	1
2	2	1	3	7	6	5	4
3	1	2	3	6	7	4	5
4	6	5	7	3	4	2	1
5	4	5	3	2	1	7	6
6	2	1	3	4	5	7	6
7	6	7	5	1	2	4	3

distribution, which incidentally also provides a constructive proof of the existence and uniqueness of this distribution, proceeds as follows:

(i)   the origins are first assigned their top rated destinations;

(ii)   the line of destination numbers is checked to see whether any number is repeated—if not, the origin-optimal distribution has been reached—if there are repeats, proceed to (iii);

(iii)   take any pair of repeated numbers and from the destination preferences retain the number which has the higher preference and change the other number to the number corresponding to the next lowest destination preference;

(iv)   return to Step (ii).

Thus, (i) gives the first line in Table 23.3. Since 7 is repeated, the preference ordering for destination 7 is checked from Table 23.2, and origin 5 has a higher rating than origin 3, so the trip from origin 5 to destination 7 is retained. The destination preferences for origin 3 are checked and instead of destination 7, destination 6, which is the next lowest, is assigned. This gives the second line of Table 23.3. The procedure is repeated until the final line in the table is reached, signifying

TABLE 23.3 ORIGIN-OPTIMAL DISTRIBUTION

			Origins			
1	2	3	4	5	6	7
5	1	7	4	7	3	6
5	1	6	4	7	3	6
5	1	6	4	7	3	7
5	1	6	4	7	3	5
5	1	6	4	7	3	4
5	1	6	5	7	3	4
4	1	6	5	7	3	4
2	1	6	5	7	3	4

all destinations assigned and the origin-optimal destination. It is instructive to check that this distribution is stable. For example, using the notation that origin i is paired with destination $s(i), s(1) = 2$ and $s(7) = 4$. Origin $i = 1$ prefers destination 4 to destination 2, but destination 4 prefers origin 7 to origin 1.

To apply this procedure to the practical problem of trip distribution, it is necessary to group the origins and destinations into zones and replace individual trip preferences by group preferences. For calibration purposes, the group preferences are determined from opportunity curves. By an extension of the algorithm described above, an origin-optimal or a destination-optimal zone-to-zone trip distribution is obtained. The number of trips f_{ij} from origin zone i to destination zone j is the total number of individual trips originating from the group a_i origins in zone i which have destinations within the group of b_j destinations in zone j.

The trip preferencing model automatically yields nonnegative integral trip numbers which are correctly balanced, satisfying the conservation equations. The model is neither compressible nor separable. The reader is referred to [10] for an account of the application of the model to actual data.

24. Combined Distribution and Assignment

It is customary in the transportation planning process to consider traffic assignment separately from trip distribution, but it is important to realize that the two are dependent on each other. The deterrence function used in trip distribution models depends on the congestion and traffic flow on the networks, and the output of the assignment model should serve as an input to the distribution model. This feedback is difficult to incorporate into the planning process, but its neglect is likely to be highly significant.

Several attempts have been made to combine distribution with assignment, and some of the methods which have been developed will be discussed below. The simplest resort to repeated use of distribution and assignment programs.

(a) TRC Program

For a transportation study of Toronto, Canada, Traffic Research Corporation [16] developed an iterative procedure which specifically allowed for the feedback between distribution and assignment on the basis of an assumed cost-flow relation. At the end of the first pass of the process, the link flows were obtained by all-or-nothing assignment and the link costs updated. For the second iteration, new cheapest route trees were built, with new interzonal costs determined for input to the distribution model. The O–D traffic was then assigned in some proportion to the two cheapest routes already determined. The whole iteration was repeated until convergence was obtained. The program allowed up to 9 routes between each O–D pair.

(b) LTS Program

The analytical methods developed for the London Transportation Study [17] allowed for feedback between congested assignment and trip distribution as follows. All-or-nothing assignment was first used to indicate where the traffic demand exceeded the network capacity. The speeds which had been assumed for the overloaded links were then reduced until all roads were equally overloaded. Two successive speed changes were needed to achieve this balance, and each time the speeds were altered, the trips were redistributed.

A more drastic technique achieved a reduction in trips in order to accommodate the traffic. This was assumed to correspond to some form of restraint on trips by regulation or pricing.

The multiple route assignment described in Sect. 14(d), Chap. III, has also been used in an iterative distribution-assignment scheme. Its application in the LTS has indicated that it forces a fairly rapid convergence and that, except when the initial overloading of the network is severe, redistribution of the trips is only necessary on alternate passes.

(c) Multicommodity Distribution–Assignment

From the theoretical point of view, the most satisfactory method of combining distribution with assignment is to incorporate them in the one minimum network cost multicommodity flow model. The combined model requires some extension of notation.

We denote centroids by $k = 1, 2, ..., n$, and an origin–destination pair by k, k'. The trip productions and attractions, which usually form the input to the distribution models, are denoted by a^k and $b^{k'}$. The flow of commodity k' on the jth chain from k to k' is denoted by $h_j^{kk'}$ with corresponding route cost $C_j^{kk'}$. The minimum network cost problem is then represented by the program

$$h_j^{kk'} \geqslant 0, \tag{1}$$

$$\sum_{k'} \sum_j h_j^{kk'} = a^k, \tag{2}$$

$$\sum_k \sum_j h_j^{kk'} = b^{k'}, \tag{3}$$

$$\sum_k \sum_{k'} \sum_j C_j^{kk'} h_j^{kk'} = C(\text{min}). \tag{4}$$

Link capacities can be introduced if required.

Although this model is attractive from the theoretical point of view, its practical use is limited because of the difficulties in solving the mathematical program for large networks with flow dependent link costs. Nevertheless, some interesting calculations have been carried out and reported [18].

25. Conclusion

In describing various trip distributions, the emphasis has been placed on their basic structures and properties. It has been shown where the extremal principles analyzed in the previous chapter are inherent in the models. Except for the models which are widely used and are well documented with users' manuals, no attempt has been made to give practical details of how the models should be programmed for computer analysis.

Transportation planners are concerned with the choice of the distribution model best fitted to their purposes. Comparative evaluations of the different models have been made by various authors [12], [13] and these together with other critical reviews [8], [14] tend to be rather inconclusive and underline the necessity for a better understanding of the model structures. An interesting comparison of the Hitchcock and gravity models has been made for a theoretical town [15].

The preferencing model has been described in some detail because of its novelty. The present trend in research tends to favor the further analysis of opportunity curves because of their usefulness in calibration and prediction.

26. Notes and References

[1] Hitchcock, F. L., The Distribution of a Product from Several Sources to Numerous Localities, *J. Math. and Phys.*, **20**, 224–280 (1941).

Because of the rather confused history of the development of linear programming and its applications, it is probably wisest to avoid attaching specific authors' names to models and problems, but the transportation problem is often called the Hitchcock problem although the double barrelled Hitchcock–Koopmans epithet is preferred by some. This early paper clearly formulates the model. Although the given solution is somewhat sketchy, its simplicity is still attractive and provides an excellent introduction to the classical transportation problem.

[2] Murchland, J. D., Some Remarks on the Gravity Model of Traffic Distribution and an Equivalent Maximizing Formulation, *Report LSE-TNT-38* (1966).

This short excellent report contains some interesting comments on the gravity model and its formulation as a maximization problem.

[3] Wilson, A. G., The Use of the Concept of Entropy in System Modelling, *Operations Res. Quart.*, **21**, 247–265 (1970).

Some controversy has arisen over the use of the concept of entropy in the social sciences. The point of view expounded in this paper is challenged by D. J. White in a short note on pp. 279–281 of the same journal. The point at issue is whether entropy can be used to predict equilibrium distribution patterns in the context of travellers who know where they want to go and select rational alternatives for getting there. It is agreed by various authors that the entropy concept is helpful in formulating trip distribution models—but why it works is the question. M. J. Beckman and T. F. Golob, argue in a paper (On the Metaphysical Foundations of Traffic Theory: Entropy Revisited, presented at the Fifth International Symposium on the Theory of Traffic Flow and Transportation, held at the University of California, Berkeley, in June, 1971) that the trip distribution formulae are better derived from a utility maximization model for which the basic assumption is that "households are assumed to behave rationally (i.e., to maximize net utility)." The paper by Wilson includes an extensive list of references, among them Vol. **14**, No. 1, (1970) of *Transportation Res.*, an issue devoted solely to trip distribution models.

[4] Sasaki, T., Probabilistic Models for Trip Distribution, Proceedings of the Fourth International Symposium on the Theory of Traffic Flow, Karlsruhe (1968).

Professor Sasaki has written a series of papers on the application of Markov chain theory to trip distribution, and the present paper exploits the entropy maximization method. The distribution model is applied, with numerical results, to shopping trips in Kyoto City and to the trip distribution between ramps on the Hanshin Expressway.

[5] Tomlin, J. A. and Tomlin, S. G., Traffic Distribution and Entropy, *Nature*, **220**, 974–976 (1968).

This son–father paper uses the theory of statistical mechanics to formulate distribution models in terms of the entropy concept. A model analogous to the fixed mean trip length model is described and checked with real data from the Metropolitan Adelaide Transportation Study. A good fit to the data is obtained with $\gamma = 1/c$, in the notation of (66). Sect. 22.

[6] Sinkhorn, R., Diagonal Equivalence to Matrices with Prescribed Row and Column Sums, *Amer. Math. Monthly*, **74**, 402–405 (1967).

The author proves the following:

"THEOREM: *Let* $r_1, \ldots, r_m, c_1, \ldots, c_n$ *be fixed positive numbers. Then, corresponding to each positive* $m \times n$ *matrix* **A**, *there is a unique matrix* *of the form* $\mathbf{D}_1 \mathbf{A} \mathbf{D}_2$ *with row sums* $\mu r_1, \ldots, \mu r_m$ *and column sums* $c_1, \ldots, c_n$ *where* $\mu = \sum_j c_j / \sum_i r_i$. $\mathbf{D}_1$ *and* $\mathbf{D}_2$ *are respectively* $m \times m$ *and* $n \times n$ *diagonal matrices with positive diagonals and are themselves unique up to a scalar multiple.*

The iterative process of alternately scaling the rows and columns of $\mathbf{A}$ *to have row and column sums respectively* r_i *and* c_j *can be used to find* $\mathbf{D}_1 \mathbf{A} \mathbf{D}_2$. *The subsequence from the iteration, in which column sums are scaled, converges to* $\mathbf{D}_1 \mathbf{A} \mathbf{D}_2$ *while the subsequence in which the row sums are scaled converges to* $(1/\mu) \mathbf{D}_1 \mathbf{A} \mathbf{D}_2$. *In particular if* $\sum_i r_i = \sum_j c_j$ *then the entire iteration converges to* $\mathbf{D}_1 \mathbf{A} \mathbf{D}_2$.

In an earlier paper to which the author refers, examples are given to illustrate the possible breakdown of the iterative process when the matrix $\mathbf{A}$ has zero elements. It is also shown that replacing the zero elements by small positive quantities, iterating, and then taking the limit does not lead to unique results.

[7] Tanner, J. C., Factors Affecting the Amount of Travel, *Road Res. Lab. Tech. Paper*, **51**, Her Majesty's Stationery Office, London (1961).

In this important paper, Tanner discusses possible deterrence functions in considerable detail. He points out that when using t (time or distance) as a continuous variable, so that sums become integrals, a deterrence function of the form t^{-n} theoretically implies infinite total travel time on long trips if $n < 3$, and infinite total travel time on short trips if $n \geqslant 3$. The singular behavior of the integrals can be eliminated by choice of a function of the form $t^{-b} e^{-at}$ corresponding to $\gamma(t) = at + b \ln t$ in the notation of (86), Sect. 22.

[8] Heggie, I. G., Are Gravity and Interactance Models a Valid Technique for Planning Regional Transport Facilities?, *Operations Res. Quart.*, **20**, 93–110 (1969).

This paper concisely defines the gravity and interactance distribution models and then asks the penetrating questions: are the hypotheses reasonable; are the models logically consistent; and do the models fit the facts? The author's general conclusion from his analysis is that gravity and interactance models do not provide a valid means of producing traffic forecasts in a regional development. In "A Rejoinder" on pp. 489–492 of the same journal volume, A. G. Wilson criticizes Heggie's paper as offering "an extremely misleading review of gravity and interactance models." Wilson's note is followed by a spirited reply from Heggie—only to be followed by "A Further Rejoinder" by Wilson. The lively debate on this controversy gives an interesting insight into the gravity models and their applications.

[9]   Stouffer, S. A., Intervening Opportunities: A Theory Relating Mobility and Distance, *Amer. Sociol. Rev.*, **5**, 845–867 (1940).

This interesting and readable paper propounds the theory that there is no necessary relationship between the mobility of migrating people and the distance they migrate, but that "the number of persons going a given distance is directly proportional to the number of opportunities at that distance and inversely proportional to the number of intervening opportunities." This important concept of opportunities is defined by the author in terms of the particular problem being analyzed. Thus "for a white family leaving a dwelling in rental group K in tract X, the number of opportunities in tract Y is proportional to the total number of white families, whatever their place of origin, moving to dwellings in rental group K within tract Y." The paper includes an extensive test of the proposed theory with empirical data.

[10]   Kirby, R. F., A Preferencing Model for Trip Distribution, *Transportation Sci.*, **4**, 1–35 (1970).

This paper gives a detailed account of the theory of the trip preferencing model and a formulation of the algorithm for its practical application. The model is illustrated with an example using data from a transportation study of the city of Launceston, Australia. The paper is a condensed version of part of the author's doctoral thesis.

[11]   Gale, D. and Shapley, L. S., College Admissions and the Stability of Marriage, *Amer. Math. Monthly*, **69**, 9–15 (1962).

This entertaining paper describes in a particularly lucid manner the basic ideas of a preference model for the distribution, not of trips, but of students to colleges. The model is also applied to the problem of determining stable and optimal marriages in a community! Although the structure of the preference model is easy to describe in a loose nonmathematical way, the precise mathematical formulation of an algorithm (given in the previous reference) is quite complicated.

[12]   Heanue, K. E. and Pyers, C. E., A Comparative Evaluation of Trip Distribution Procedures, *Public Roads*, **34**, 43–51 (1966).

This important paper reports on a research project designed to test and evaluate the following trip distribution models: Fratar, gravity, intervening opportunities, and competing opportunities. The validity of the models for forecasting purposes is analyzed by comparison with a seven-year historical period for Washington, D.C. The project concluded by rating the gravity and intervening opportunities models about equal and somewhat better than the other two.

[13]   Lawson, M. C. and Dearinger, J. A., A Comparison of Four Work Trip Distribution Models, *J. Highway Div., Proc. Amer. Soc. Civil Eng.*, **93**, 1–25 (1967).

This paper reports the results of a comparison of four trip distribution models: electrostatic, gravity, competing opportunities, and multiple regression. The comparison is made for work trips in Lexington, Kentucky. Various suggestions are made for improvements to the models although the gravity model is favored as giving the best results.

[14]   Fairthorne, D. B., Description and Shortcomings of Some Urban Road Traffic Models, *Operations Res. Quart.*, **15**, 17–28 (1964).

This paper gives a lucid description of the gravity and opportunity models and is particularly valuable for its analysis of the theoretical bases of these models. The author, while admitting that the ultimate test of a model is the accuracy of its predictive power in practical applications, suggests that a useful practical model is likely also to be one which is theoretically consistent. The inconsistencies and shortcomings of the gravity and opportunity models are revealed in some detail.

[15]   McDonald, W. R. and Blunden, W. R., The Application of Linear Programming to the Determination of Road Traffic Desire Line Patterns, *Proc. Australian Road Res. Board Conf.*, **4**, 153–168 (1968).

This paper compares the Hitchcock trip distribution model (referred to simply as the linear programming model) with the gravity model for the hypothetical city Lautsville, developed by the Los Angeles Regional Transportation Study as a test city for illustrating basic principles involved in forecasting travel. The authors give cogent reasons for the usefulness of the Hitchcock model, especially for planning purposes.

[16]   Irwin, N. A. and Von Cube, A. G., Capacity Restraint in Multi-Travel Mode Assignment Programs, *H.R.B. Bull.*, **347**, 258–289 (1962).

The TRC programs have been noteworthy for their consistent and valid approach to transportation planning. This paper describes the basic program blocks—tree generation, time factor, trip distribution, proportional split, assignment and link updating—and explains the interconnection between these. The incorporation of proportional (modal) split within the iterations is described in detail.

[17] Tresidder, J. O., Meyers, D. A., Burrell, J. E. and Powell, T. J., The London Transportation Study: Methods and Techniques, *Proc. Inst. Civil Eng.*, **39**, 433–464 (1968).

This paper is an excellent summary of the new improved techniques which were developed for the LTS. As can be expected for a conurbation of the size and complexity of the London Study Area, the techniques are principally designed to allow the estimation of flows on a heavily loaded transportation network, with the emphasis on the economic evaluation of alternative transportation plans.

[18] Tomlin, J. A., A Mathematical Programming Model for the Combined Distribution-assignment of Traffic, *Transportation Sci.*, **5**, 122–140 (1971).

The author illustrates his theory by applying his method to the network used by Charnes and Cooper in [24], Sect. 17, Chap. III.

27. Problems

1. A traffic desire network has 3 origins, $i = 1, 2, 3$, and 4 destinations, $j = 1, 2, 3, 4$, and the number of trips a_i originating from i, the number of trips b_j with destination j, and the costs c_{ij} of a single trip from i to j are given by the table

	1	2	3	4	a_i
1	5	4	3	2	5
2	10	8	4	7	5
3	9	9	8	4	5
b_j	1	6	2	6	$15 = v$.

The solution of the Hitchcock trip distribution model gives the following optimal trip table with the associated implicit prices

$$\alpha_i$$

$$
\begin{array}{cccc|c}
1 & 3 & 0 & 1 & 5 \\
0 & 3 & 2 & 0 & 9 \\
0 & 0 & 0 & 5 & 7 \\
\end{array}
$$

$$\beta_j \quad 0 \quad -1 \quad -5 \quad -3 \ .$$

If destination nodes $j = 3, 4$ are combined, calculate the new costs according to (11)–(13), Sect. 21, and verify that the compressed trip table is optimal.

2. Repeat Problem 1 but with destinations 1 and 2 combined instead of destinations 3 and 4.

3. A traffic desire network with four centroids has productions and attractions as follows:

Centroid	1	2	3	4
Production	40	30	20	10
Attraction	10	20	30	40

Calculate the trip table for the proportional distribution model. Verify the compressibility property by combining centroids 2 and 3 and the separability property by removing centroid 2.

4. For the same data as in Problem 1, use the iterative procedure described in Sect. 22 to determine the trip table giving a mean trip length of 5.50 measured in the same units as c_{ij}.

5. For a study area with 3 zones, the trip table for the base year is measured to be

$$a_i^{(1)}$$

$$
[f_{ij}^{(1)}] = \begin{bmatrix} 1 & 4 & 2 \\ 6 & 2 & 3 \\ 4 & 1 & 2 \end{bmatrix} \begin{matrix} 7 \\ 11 \\ 7 \end{matrix}
$$

$$b_j^{(1)} \quad 11 \quad 7 \quad 7 \quad 25 = v \ .$$

It is predicted that the growth factors to give the estimated productions and attractions for the design year are

Growth factors

Zone	Production	Attraction
1	2	3
2	3	4
3	4	2 .

Use a matrix scaling method to estimate the trip table for the design year.

6. A traffic desire network with 3 centroids has productions and attractions as follows:

Centroid	1	2	3
Production	14	33	28
Attraction	33	28	14 .

The inter- and intrazonal travel times as given by a skim trees program are

$$[c_{ij}] = \begin{matrix} & 1 & 2 & 3 \\ 1 & \begin{bmatrix} 8 & 1 & 4 \\ 2 & 3 & 6 & 4 \\ 3 & 2 & 7 & 4 \end{bmatrix} \end{matrix}$$

and the travel times factors F_{ij} are given by the following table:

c_{ij}	1	2	3	4	5	6	7	8
F_{ij}	82	52	50	41	39	26	20	13 .

Assuming K-factors are all unity, use the scaling technique of the gravity model to obtain three iterations of the trip distribution table.

7. A traffic desire network with 3 centroids has productions and attractions as in Problem 6, but the zone-to-zone costs are

$$[c_{ij}] = \begin{bmatrix} 1 & 2 & 3 \\ 2 & 1 & 3 \\ 2 & 3 & 1 \end{bmatrix} .$$

The L-factors for the 3 zones are $L_1 = L_3 = 0.04, L_2 = 0.02$. Use the intervening opportunities model to determine the distribution of trips.

8. For the same data as given in Tables 23.1 and 23.2, determine the destination-optimal trip distribution.

9. (a) Find the (normalized) flows ρ_{ij}, $i = 1, ..., m, j = 1, ..., n$, which maximize the network entropy

$$H = -\sum_{i,j} \rho_{ij} \ln \rho_{ij},$$

subject to

   (i) $\sum_{i,j} \rho_{ij} = 1$;

   (ii) $\sum_{j} \rho_{ij} = u_i$;

   (iii) $\sum_{i} \rho_{ij} = v_j$;

   (iv) $\rho_{11} = k$.

where u_i, v_j and k are given positive quantities.

   (b) Find a numerical solution for the case where $m = n = 3$, $u_1 = u_2 = 0.25$, $u_3 = 0.50$, $v_1 = v_3 = 0.40$, $v_2 = 0.20$ and $\rho_{11} = 0.01$.

A

THEOREM FOR CHEAPEST ROUTE ALGORITHMS

In this appendix we prove the following

THEOREM: *The costs of the cheapest routes from node n_1 to nodes n_r of a network $[N; L]$ with positive link costs $c(n_i, n_j)$ are the unique solutions of the functional equations*

$$f(n_1) = 0, \tag{1}$$

$$f(n_r) = \min_{n_i \neq n_r} [f(n_i) + c(n_i, n_r)], \qquad n_r \neq n_1. \tag{2}$$

Proof: For simplicity, we suppose that the network is connected and undirected, and we adopt the usual convention that $c(n_i, n_j) = \infty$ if $(n_i, n_j) \notin L$. Denote the cost of a path from n_1 to n_r by $C(n_1, n_r)$, and the cost of a cheapest path from n_1 to n_r by $C^*(n_1, n_r)$. Then we have to prove that $C^*(n_1, n_r)$, for given n_1, and for all $n_r \in N$, are the unique solutions of (1) and (2).

(a) The first step in the proof is to show that the cheapest costs $C^*(n_1, n_r)$ satisfy (1) and (2). The cheapest path from n_1 to n_1 is the "empty" path without links for which $C^*(n_1, n_1) = 0$, so that (1) is satisfied. Suppose $n_r \neq n_1$ and let $n_1, \ldots, n_j, n_r$ be a cheapest path with cost $C^*(n_1, n_r)$.

Then

$$C^*(n_1, n_r) = C(n_1, n_j) + c(n_j, n_r), \tag{3}$$

where $C(n_1, n_j)$ is the cost of the path $n_1, \ldots, n_j$. Hence,

$$C^*(n_1, n_r) \geqslant C^*(n_1, n_r) + c(n_j, n_r) \tag{4}$$

$$\geqslant \min_{n_i \neq n_r} [C^*(n_1, n_i) + c(n_i, n_r)] \tag{5}$$

$$= C^*(n_1, n_j') + c(n_j', n_r). \tag{6}$$

Relation (4) follows from (3) because $C \geqslant C^*$, and in (6) n_j' is a node $\neq n_r$, for which the minimum in (5) is attained.

Attention is now turned to a path $n_1, \ldots, n_i', \ldots, n_j'$ which has cost $C^*(n_1, n_j')$. We extend this path to give a route $n_1, \ldots, n_i', \ldots, n_j', n_r$ from n_1 to n_r, and we prove that this route is a path by showing that it does not visit node n_r twice. If the contrary were true, so that $n_i' = n_r$, say, then $n_1, \ldots, n_i' = n_r$ is a path from n_1 to n_r with cost

$$C'(n_1, n_i') < C^*(n_1, n_j') + c(n_j', n_r) \tag{7}$$

$$\leqslant C^*(n_1, n_r), \tag{8}$$

by (4)–(6). This implies $n_1, \ldots, n_i'$ is a path from n_1 to n_r with cost less than $C^*(n_1, n_r)$, which is impossible. Thus, $n_1, \ldots, n_i', \ldots, n_j', n_r$ is indeed a path from n_1 to n_r, and its cost is

$$C'(n_1, n_r) = C^*(n_1, n_j') + c(n_j', n_r) \tag{9}$$

$$\leqslant C^*(n_1, n_r), \tag{10}$$

by (4)–(6). But since $C^*(n_1, n_r)$ is the cheapest cost, the $\geqslant$ signs in (4) and (5) are equal signs. Inequality (5), as it now becomes, proves that the C^* satisfy (2). Note also that (4) with an equal sign now implies that $n_1, \ldots, n_j$ is a cheapest path from n_1 to n_j.

(b) We next prove that the solutions to (1) and (2) define a path from n_1 to n_r. We first observe that if $f(n_r)$, $n_r \in N$, is any solution set of (1) and (2), then $0 < f(n_r) < \infty$ if $n_r \neq n_1$.

   We also note that if the minimum in (2) is attained for a node n_{r-1} (not necessarily unique), then

$$f(n_r) = f(n_{r-1}) + c(n_{r-1}, n_r). \tag{11}$$

We call n_{r-1} a predecessor node of n_r.

We next show that for given n_r, the solution set of (1) and (2) defines a path $n_1, n_2, ..., n_{k-1}, n_k, ..., n_r$ from n_1 to n_r, such that

$$f(n_k) - f(n_{k-1}) = c(n_{k-1}, n_k), \qquad k = 2, 3, ..., r. \tag{12}$$

As noted above, the solution of (2) defines a predecessor node n_{r-1} of n_r and successively a sequence of nodes $n_r, n_{r-1}, ..., n_k, n_{k-1}, ...$ with n_{k-1} the predecessor of n_k. From (11), the relation (12) holds for each pair of successive nodes. It is only necessary to show that the sequence of nodes defines a path from n_1 to n_r. This is a consequence of the fact that $f(n_r), f(n_{r-1}), ..., f(n_k), f(n_{k-1}), ...$, is a strictly decreasing sequence, which negates the possibility that a node can appear more than once in the sequence, and forces the sequence to reach n_1, the only node for which $f(n_i) = 0$.

(c) Finally, we show that the solutions to (1) and (2) are unique by supposing, to the contrary, that f_1 and f_2 are two solution sets and that for some $n_r \in L, f_1(n_r) < f_2(n_r)$. Since $f_1(n_1) = f_2(n_1) = 0$, n_r is different from n_1. We suppose that $n_1, ..., n_{r-1}, n_r$ is a path from n_1 to n_r, such that, as shown in (12),

$$f_1(n_k) - f_1(n_{k-1}) = c(n_{k-1}, n_k), \qquad k = 2, 3, ..., r. \tag{13}$$

In particular,

$$\begin{aligned} f_1(n_r) &= f_1(n_{r-1}) + c(n_{r-1}, n_r) \\ &< f_2(n_r) \leqslant f_2(n_{r-1}) + c(n_{r-1}, n_r), \end{aligned} \tag{14}$$

since $f_2(n_r)$ is a solution of (2). From (14) it follows that $f_1(n_{r-1}) < f_2(n_{r-1})$, and the argument is continued until we deduce that $f_1(n_1) < f_2(n_1)$, which is impossible since both are equal to zero. The uniqueness of the solution set of (1) and (2) is thus demonstrated.

The theorem has been proved for positive link costs. If negative link costs are permitted, then the above proof can be modified to show that a necessary and sufficient condition for the theorem to be valid is that the sum of the link costs around any mesh in the network be nonnegative.

B

DUALITY THEORY

With notation which is unrelated to that used in the text, a primal linear program and its dual can be written in general form as

Primal P Dual D

$$x_j \geqslant 0, \qquad \sum_1^m y_i a_{ij} \leqslant c_j, \qquad j = 1, ..., n_1, \quad (1)$$

$$x_j \text{ unrestricted} \quad \sum_1^m y_i a_{ij} = c_j, \qquad j = n_1 + 1, ..., n, \atop \text{in sign,}$$

$$(2)$$

$$\sum_1^n a_{ij} x_j \geqslant b_i, \qquad y_i \geqslant 0, \qquad i = 1, ..., m_1, \quad (3)$$

$$\sum_1^n a_{ij} x_j = b_i, \qquad y_i \text{ unrestricted} \quad i = m_1 + 1, ..., m, \atop \text{in sign,}$$

$$(4)$$

$$\sum_1^n c_j x_j = z(\min), \qquad \sum_1^m b_i y_i = v(\max). \qquad (5)$$

The main results of duality theory are:

(a) the dual of D is P;

(b) if x_j is a feasible solution of P and y_i is a feasible solution of D (so that the constraints (1)–(4) are satisfied), then

$$z = \sum_1^n c_j x_j \geqslant v = \sum_1^m b_i y_i; \tag{6}$$

(c) if P and D both have feasible solutions, then they have optimal solutions $x_j{}^*$ and $y_i{}^*$, and the optimal values of z and v are equal:

$$z^* = v^*. \tag{7}$$

For optimal solutions, the following complementary slackness relations are valid:

(d) if $\quad x_j{}^* > 0,\quad$ then $\sum_1^m y_i{}^* a_{ij} = c_j,\ \ j = 1, \ldots, n_1,$

$$\tag{8}$$

(e) if $\sum_1^m y_i{}^* a_{ij} < c_j,\quad$ then $\quad x_j{}^* = 0,\quad j = 1, \ldots, n_1,$

$$\tag{9}$$

(f) if $\quad y_i{}^* > 0,\quad$ then $\sum_1^n a_{ij} x_j{}^* = b_i,\quad i = 1, \ldots, m_1,$

$$\tag{10}$$

(g) if $\sum_1^n a_{ij} x_j{}^* > b_i,\quad$ then $\quad y_i{}^* = 0,\quad i = 1, \ldots, m_1.$

$$\tag{11}$$

C

INEQUALITIES FOR MARGINAL AND AVERAGE LINK AND CHAIN COSTS

Throughout this book we have used the idea of the average cost of link flow $c_i(f_i)$, the total cost of a link flow $f_i c_i(f_i)$, the marginal cost of link flow $d_i(f_i)$, and corresponding average and marginal costs

$$C_j^{(k)} = C_j^{(k)}(\mathbf{h}) = \sum_i c_i(f_i) a_{ij}^{(k)}, \qquad j \in M^{(k)}$$

$$j = 1, 2, \dots, m(k), \qquad (1)$$

$$D_j^{(k)} = D_j^{(k)}(\mathbf{h}) = \sum_i d_i(f_i) a_{ij}^{(k)},$$

for chains connecting the $O^{(k)}$–$D^{(k)}$ pair. The quantity $a_{ij}^{(k)}$ is unity if link i is a member of chain $m_j^{(k)}$ connecting $O^{(k)}$–$D^{(k)}$, zero otherwise.

The purpose of this appendix is to derive certain inequalities for average and marginal costs of link and chain flows given that for the total cost of link flow:

(i) $f_i c_i(f_i)$ is strictly convex and increasing in $[0, \infty)$.

(ii) $\displaystyle \lim_{f_i \to 0^+} f_i c_i(f_i) = 0.$ \hspace{3cm} (2)

It should be pointed out that condition (i) implies that $f_i c_i(f_i)$ is continuous ([23], Sect. 17, Chap. III). In the remainder of this Appendix, and in Sect. 16, we use the following result.

159

If $0 \leqslant x_1 < x_2 < x_3$ are three values of x, and $y_1 = y(x_1)$, $y_2 = y(x_2)$, $y_3 = y(x_3)$ are the three corresponding values for the strictly convex function $y(x)$, then the following inequalities hold:

$$\frac{y_2 - y_1}{x_2 - x_1} < \frac{y_3 - y_1}{x_3 - x_1} < \frac{y_3 - y_2}{x_3 - x_2}. \tag{3}$$

In other words, the slopes of the three chords connecting any pair of the three points $(x_1, y_1)(x_2, y_2)$ and (x_3, y_3) satisfy (3). Since $y(x)$ is strictly convex, we have from its definition

$$\alpha y(z_1) + (1 - \alpha) y(z_2) > y(\alpha z_1 + (1 - \alpha) z_2), \tag{4}$$

for any $z_2 \neq z_1$ and every α such that $0 < \alpha < 1$.

Consider first the chord connecting (x_1, y_1) with (x_3, y_3) and the chord connecting (x_1, y_1) with (x_2, y_2). Since $x_1 < x_2 < x_3$, there exists a value of α, say $\bar{\alpha}$, such that

$$x_2 = \bar{\alpha} x_1 + (1 - \bar{\alpha}) x_3. \tag{5}$$

From the definition of a strictly convex function in (4) we obtain

$$y_2 = y(x_2) < \bar{\alpha} y(x_1) + (1 - \bar{\alpha}) y(x_3) = \bar{\alpha} y_1 + (1 - \bar{\alpha}) y_3. \tag{6}$$

Therefore,

$$\frac{y_2 - y_1}{x_2 - x_1} < \frac{\bar{\alpha} y_1 + (1 - \bar{\alpha}) y_3 - y_1}{\bar{\alpha} x_1 + (1 - \bar{\alpha}) x_3 - x_1} = \frac{(y_3 - y_1)}{(x_3 - x_1)}. \tag{7}$$

It is also easy to show that the right-hand inequality of (3) holds. We can now show that conditions (i) and (ii) on the total cost of flow imply that the average cost of flow, $c_i(f_i)$, is monotone increasing.

We use the three chords lemma in (3) with $y = f_i c_i(f_i)$. Substituting $x_1 = 0$, $x_2 = f_i$, $x_3 = f_i'$ into (3), yields the inequality

$$\frac{f_i c_i(f_i)}{f_i} < \frac{f_i' c_i(f_i')}{f_i'},$$

or

$$c_i(f_i) < c_i(f_i'), \qquad \text{for} \quad 0 < f_i < f_i'. \tag{8}$$

Thus, we have shown that conditions (i) and (ii) in (2) imply that $c_i(f_i)$ is monotone increasing.

To obtain inequalities for marginal costs of link and chain flows, we make use of the well-known mean-value theorem ([23], Sect. 17, Chap. III):

if $y(x)$ is a differentiable function of x in the open interval (a, b), then there always exists a particular value of x, say z, such that $a < z < b$, and

$$y(b) - y(a) = (b - a) \frac{dy(x)}{dx}\bigg|_{x=z}. \tag{9}$$

Geometrically, the theorem states that one can always find a value of x for which the slope (tangent line) at x times an interval of length $b - a$ is exactly equal to the difference in the values of the function evaluated at a and b, respectively. The only requirement on z is that it lies between a and b.

In conjunction with properties (i) and (ii), (9) can be used to prove that the average cost of link flow $c_i(f_i)$ is strictly less than the marginal cost $d_i(f_i)$, and that the following inequalities also hold:

$$f_i' c_i(f_i') - f_i c_i(f_i) > d_i(f_i)(f_i' - f_i), \tag{10}$$
$$f_i' c_i(f_i') - f_i c_i(f_i) < d_i(f_i')(f_i' - f_i). \qquad f_i' \neq f_i, \tag{11}$$

The difference in the two inequalities is simply that the inequality is reversed when the marginal link cost is evaluated at f_i' rather than f_i. We know, from (9), that if $f_i < f_i'$, there exists some value $\bar{f}_i$, $f_i < \bar{f}_i < f_i'$, for which

$$f_i' c_i(f_i') - f_i c_i(f_i) = d_i(\bar{f}_i)(f_i' - f_i). \tag{12}$$

Making use of the fact that for a strictly convex total cost the marginal costs are strictly increasing with flow,

$$d_i(f_i) < d_i(\bar{f}_i) < d_i(f_i'), \tag{13}$$

we are led to the inequalities of (10) and (11) for $f_i < f_i'$. A similar proof leads to the same inequalities when $f_i > f_i'$. A simple sketch will illustrate the result.

It is a simple matter to show that these inequalities carry over to average and marginal costs of chain flows. By summing (8) over all links in a given chain, we obtain

$$C_j^{(k)}(\mathbf{h}') = \sum_i c_i(f_i') a_{ij}^{(k)} > \sum_i c_i(f_i) a_{ij}^{(k)} = C_j^{(k)}(\mathbf{h}), \tag{14}$$

for any chain where at least one link flow $f_i' > f_i$ and all other link flows in the chain do not decrease. By $\mathbf{h}'$, we mean the flow pattern corresponding to the new link flow vector $\mathbf{f}'$, i.e.,

$$f_i' = \sum_k \sum_j a_{ij}^{(k)} h_j'^{(k)}. \tag{15}$$

As an example, if the flow in one chain changes from $h_j^{(k)}$ to

$$h_j'^{(k)} = h_j^{(k)} + \Delta, \qquad \Delta > 0, \tag{16}$$

then at least one link in that chain has an increase of flow equal to Δ (recall that by our convention in Sect. 7, Chap. II, two distinct chains connecting an $O^{(k)}$–$D^{(k)}$ pair must have at least one distinct link).

We can also obtain similar inequalities for marginal chain costs. Substituting (10) and (11) into (1) gives

$$D_j^{(k)}(\mathbf{h})\Delta = \sum_i d_i(f_i) a_{ij}^{(k)} \Delta$$

$$< \sum_i [f_i' c_i(f_i') - f_i c_i(f_i)] a_{ij}^{(k)}, \qquad \Delta > 0, \tag{17}$$

where $h_j'^{(k)} = h_j^{(k)} + \Delta$, and in at least one link $f_i' = f_i + \Delta$. Relation (17) simply says that the difference in total costs of all links in the jth chain (due to the new flows) is greater than the marginal chain cost times the increase in flow. The inequality is reversed if either Δ is negative or the marginal link costs are evaluated at $f_i + \Delta$; the inequality holds if the marginal costs are evaluated at $f_i - \Delta$.

D

ANSWERS TO PROBLEMS

Chapter II

12.1   No. *L* does not contain $(1, 4)$ or $(4, 1)$, for example.

12.2   Yes.

12.3   There are nine chains, listed below as sequences of links:

$$m_1 = (1, 2), (2, 3), (3, 4), (4, 5), (5, 6),$$

$$m_2 = (1, 2), (2, 3), (3, 4), (4, 6),$$

$$m_3 = (1, 2), (2, 3), (3, 5), (5, 6),$$

$$m_4 = (1, 2), (2, 4), (4, 5), (5, 6),$$

$$m_5 = (1, 2), (2, 4), (4, 6),$$

$$m_6 = (1, 2), (2, 5), (5, 6),$$

$$m_7 = (1, 3), (3, 4), (4, 5), (5, 6),$$

$$m_8 = (1, 3), (3, 4), (4, 6),$$

$$m_9 = (1, 3), (3, 5), (5, 6).$$

163

12.4   There are eleven paths, which are not chains. They are listed below as sequences of links:

  (i) $(1,2), (2,3), (3,5), (4,5), (4,6),$

  (ii) $(1,2), (2,4), (3,4), (3,5), (5,6),$

  (iii) $(1,2), (2,5), (3,5), (3,4), (4,6),$

  (iv) $(1,2), (2,5), (4,5), (4,6),$

  (v) $(1,3), (2,3), (2,4), (4,5), (5,6),$

  (vi) $(1,3), (2,3), (2,4), (4,6),$

  (vii) $(1,3), (2,3), (2,5), (4,5), (4,6),$

 (viii) $(1,3), (2,3), (2,5), (5,6),$

  (ix) $(1,3), (3,4), (2,4), (2,5), (5,6),$

  (x) $(1,3), (3,5), (2,5), (2,4), (4,6),$

  (xi) $(1,3), (3,5), (4,5), (4,6).$

12.5   There are seven such meshes:

  (i) $(2,3), (3,4), (2,4),$

  (ii) $(2,3), (3,4), (4,5), (2,5),$

  (iii) $(2,3), (3,5), (2,5),$

  (iv) $(2,3), (3,5), (4,5), (2,4),$

  (v) $(2,4), (3,4), (3,5), (2,5),$

  (vi) $(2,4), (4,5), (2,5),$

  (vii) $(2,4), (4,5), (3,5), (2,3).$

12.6   Spanning tree which is an arborescence:

$$(1,2), (2,3), (2,4), (4,5), (4,6).$$

Spanning tree which is not an arborescence:

$$(1,3), (2,3), (3,4), (3,5), (5,6).$$

12.7

$$\text{link } (j,k)$$

	(1,2)	(1,3)	(2,3)	(2,4)	(2,5)	(3,4)	(3,5)	(4,5)	(4,6)	(5,6)
1	1	1	0	0	0	0	0	0	0	0
2	−1	0	1	1	1	0	0	0	0	0
3	0	−1	−1	0	0	1	1	0	0	0
4	0	0	0	−1	0	−1	0	1	1	0
5	0	0	0	0	−1	0	−1	−1	0	1
6	0	0	0	0	0	0	0	0	−1	−1

$\mathbf{E} = \text{node } i$

12.8 With the chains enumerated as in the answer to Problem 12.3,

chain

	m_1	m_2	m_3	m_4	m_5	m_6	m_7	m_8	m_9
(1,2)	1	1	1	1	1	1	0	0	0
(1,3)	0	0	0	0	0	0	1	1	1
(2,3)	1	1	1	0	0	0	0	0	0
(2,4)	0	0	0	1	1	0	0	0	0
(2,5)	0	0	0	0	0	1	0	0	0
(3,4)	1	1	0	0	0	0	1	1	0
(3,5)	0	0	1	0	0	0	0	0	1
(4,5)	1	0	0	1	0	0	1	0	0
(4,6)	0	1	0	0	1	0	0	1	0
(5,6)	1	0	1	1	0	1	1	0	1

$\mathbf{A} = \text{link}$.

12.9

$$\mathbf{EA} = \begin{bmatrix} 1 & 1 & 1 & 1 & 1 & 1 & 1 & 1 & 1 \\ 0 & 0 & 0 & 0 & 0 & 0 & 0 & 0 & 0 \\ 0 & 0 & 0 & 0 & 0 & 0 & 0 & 0 & 0 \\ 0 & 0 & 0 & 0 & 0 & 0 & 0 & 0 & 0 \\ 0 & 0 & 0 & 0 & 0 & 0 & 0 & 0 & 0 \\ -1 & -1 & -1 & -1 & -1 & -1 & -1 & -1 & -1 \end{bmatrix}.$$

12.10 $g = 12$. Node 5: $f_{25}+f_{35}+f_{45} = 1+3+4 = 8 = f_{56}$. Node 6: $f_{46}+f_{56} = 4+8 = 12 = g$.

$$
\begin{bmatrix}
1 & 1 & 0 & 0 & 0 & 0 & 0 & 0 & 0 & 0 \\
-1 & 0 & 1 & 1 & 1 & 0 & 0 & 0 & 0 & 0 \\
0 & -1 & -1 & 0 & 0 & 1 & 1 & 0 & 0 & 0 \\
0 & 0 & 0 & -1 & 0 & -1 & 0 & 1 & 1 & 0 \\
0 & 0 & 0 & 0 & -1 & 0 & -1 & -1 & 0 & 1 \\
0 & 0 & 0 & 0 & 0 & 0 & 0 & 0 & -1 & -1
\end{bmatrix}
\begin{bmatrix}
5 \\ 7 \\ 0 \\ 4 \\ 1 \\ 4 \\ 3 \\ 4 \\ 4 \\ 8
\end{bmatrix}
=
\begin{bmatrix}
12 \\ 0 \\ 0 \\ 0 \\ 0 \\ -12
\end{bmatrix}.
$$

12.11 Possible chain flows corresponding to the chains as listed in the answer to Problem 12.3 are

$$h_1 = h_2 = h_3 = h_4 = 0, \qquad h_5 = 4, \qquad h_6 = 1,$$

$$h_7 = 4, \qquad h_8 = 0, \qquad h_9 = 3,$$

$$
\begin{bmatrix}
5 \\ 7 \\ 0 \\ 4 \\ 1 \\ 4 \\ 3 \\ 4 \\ 4 \\ 8
\end{bmatrix}
=
\begin{bmatrix}
1 & 1 & 1 & 1 & 1 & 1 & 0 & 0 & 0 \\
0 & 0 & 0 & 0 & 0 & 0 & 1 & 1 & 1 \\
1 & 1 & 1 & 0 & 0 & 0 & 0 & 0 & 0 \\
0 & 0 & 0 & 1 & 1 & 0 & 0 & 0 & 0 \\
0 & 0 & 0 & 0 & 0 & 1 & 0 & 0 & 0 \\
1 & 1 & 0 & 0 & 0 & 0 & 1 & 1 & 0 \\
0 & 0 & 1 & 0 & 0 & 0 & 0 & 0 & 1 \\
1 & 0 & 0 & 1 & 0 & 0 & 1 & 0 & 0 \\
0 & 1 & 0 & 0 & 1 & 0 & 0 & 1 & 0 \\
1 & 0 & 1 & 1 & 0 & 1 & 1 & 0 & 1
\end{bmatrix}
\begin{bmatrix}
0 \\ 0 \\ 0 \\ 0 \\ 4 \\ 1 \\ 4 \\ 0 \\ 3
\end{bmatrix}.
$$

The chain flows are not unique; another possible set is

$$h_1 = h_2 = h_3 = 0, \qquad h_4 = 4, \qquad h_5 = 0, \qquad h_6 = 1,$$

$$h_7 = 0, \qquad h_8 = 4, \qquad h_9 = 3.$$

12.12 $\qquad X = \{1, 2, 5\}, \qquad \overline{X} = \{3, 4, 6\},$

$$f(X, \overline{X}) = f_{13} + f_{23} + f_{24} + f_{56} = 7 + 0 + 4 + 8 = 19,$$

$$f(\overline{X}, X) = f_{35} + f_{45} = 3 + 4 = 7,$$

$$f(X, \overline{X}) - f(\overline{X}, X) = 19 - 7 = 12 = g.$$

12.13 (a) 8, for chain m_1 (see answer to Problem 12.3). (b) 7, for chain m_2. (c) $C = 102$.

12.14 There are $2^{n-2} = 2^4 = 16$ cut-sets:

X	$(X, \overline{X})$	$u(X, \overline{X})$
$\{1\}$	$\{(1,2), (1,3)\}$	14
$\{1,2\}$	$\{(1,3), (2,3), (2,4), (2,5)\}$	15
$\{1,3\}$	$\{(1,2), (3,4), (3,5)\}$	14
$\{1,4\}$	$\{(1,2), (1,3), (4,5), (4,6)\}$	26
$\{1,5\}$	$\{(1,2), (1,3), (5,6)\}$	23
$\{1,2,3\}$	$\{(2,4), (2,5), (3,4), (3,5)\}$	13
$\{1,2,4\}$	$\{(1,3), (2,3), (2,5), (4,5), (4,6)\}$	23
$\{1,2,5\}$	$\{(1,3), (2,3), (2,4), (5,6)\}$	23
$\{1,3,4\}$	$\{(1,2), (3,5), (4,5), (4,6)\}$	21
$\{1,3,5\}$	$\{(1,2), (3,4), (5,6)\}$	20
$\{1,4,5\}$	$\{(1,2), (1,3), (4,6), (5,6)\}$	31
$\{1,2,3,4\}$	$\{(2,5), (3,5), (4,6), (4,5)\}$	16
$\{1,2,3,5\}$	$\{(2,4), (3,4), (5,6)\}$	18
$\{1,2,4,5\}$	$\{(1,3), (2,3), (4,6), (5,6)\}$	27
$\{1,3,4,5\}$	$\{(1,2), (4,6), (5,6)\}$	23
$\{1,2,3,4,5\}$	$\{(4,6), (5,6)\}$	17

12.15 Feasible: $0 \leqslant f_{ij} \leqslant u_{ij}$ and $\mathbf{Ef} = \mathbf{g}$. Not maximal: $g^* = 13$.

12.16 (a) The maximal flow is 50 units obtained, for example, by a flow of 50 units on Kearny and Pacific Streets. The cut-set $(X, \overline{X}) = \{(47, 52)\}$ for $\overline{X} = \{52\}$, has cut-capacity of 50 units.

(b)   The maximal flow is 40 units obtained as follows:

nodes	chain flow
1, 7, 21, 33, 34, 35, 28, 29 30, 36, 37, 38, 47, 52	10
1, 7, 21, 33, 34, 35, 28, 23 24, 29, 30, 36, 37, 38, 47, 52	10
1, 7, 8, 9, 23, 17, 16, 24 25, 26, 38, 47, 52	10
1, 7, 8, 9, 10, 16, 11, 19 26, 38, 47, 52	10

The cut-set

$$(X, \overline{X}) = \{(16, 11), (16, 24), (23, 34), (28, 39)\}$$

for

$$\overline{X} = \{4, 5, 11, 12, 18, 19, 24, 25, 26, 27, 29, 30, 31, 36,$$
$$37, 38, 39, 40, 41, 44, 45, 46, 47, 48, 50, 51, 52\}$$

has cut capacity of 40 units.

Chapter III

18.1   Figure D.1 shows the steps in the labeling procedure and a cheapest route tree (not unique).

If the link costs are increased by 2 units, $1, 2, 4, 6$ and $1, 3, 4, 6$ are cheapest routes but $1, 2, 3, 4, 6$ is not.

18.2   The cheapest path is $1, 2, 3, 4, 6$ of cost 7 units. Paths $1, 2, 3, 4, 5, 6$; $1, 2, 4, 6$; and $1, 3, 4, 6$ are second-to-cheapest paths of cost 8 units. A general algorithm is described in Hoffman, W., and Pavley, R. A Method for the Solution of the Nth Best Path Problem, *J. Assoc. Comp. Mach.*, **6**, pp. 506–514 (1959).

18.3   The statement is false. Paths $1, 2, 4, 6, 5, 3$ and $1, 2, 5, 3$ are dearest paths of costs 14 units, but $1, 2$ is not a dearest path (in fact $1, 3, 5, 2$ and $1, 3, 5, 6, 4, 2$ are dearest paths of costs 17 units).

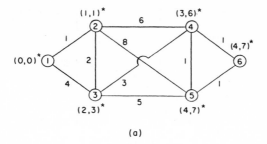

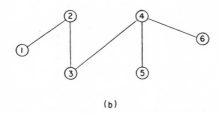

(b)

Figure D.1. Application of the tree-building algorithm. (a) Successive labels, (b) Cheapest tree.

18.4 This matrix algorithm computes at the same time the costs of cheapest routes between all points of nodes. It is stated as a 9-line ALGOL program in the following reference: Floyd, R. W., Algorithm 97, Shortest Path, *Comm. ACM*, **5**, p. 345 (1962).

The successive matrices, which are symmetric, are

$$
\mathbf{C}^{(0)} =
\begin{bmatrix}
0 & 1 & 4 & \infty & \infty & \infty \\
1 & 0 & 2 & 6 & 8 & \infty \\
4 & 2 & 0 & 3 & 5 & \infty \\
\infty & 6 & 3 & 0 & 1 & 1 \\
\infty & 8 & 5 & 1 & 0 & 1 \\
\infty & \infty & \infty & 1 & 1 & 0
\end{bmatrix},
\qquad
\mathbf{C}^{(1)} =
\begin{bmatrix}
0 & 1 & 4 & \infty & \infty & \infty \\
1 & 0 & 2 & 6 & 8 & \infty \\
4 & 2 & 0 & 3 & 5 & \infty \\
\infty & 6 & 3 & 0 & 1 & 1 \\
\infty & 8 & 5 & 1 & 0 & 1 \\
\infty & \infty & \infty & 1 & 1 & 0
\end{bmatrix},
$$

$$
C^{(2)} = \begin{bmatrix}
0 & 1 & 3 & 7 & 9 & \infty \\
1 & 0 & 2 & 6 & 8 & \infty \\
3 & 2 & 0 & 3 & 5 & \infty \\
7 & 6 & 3 & 0 & 1 & 1 \\
9 & 8 & 5 & 1 & 0 & 1 \\
\infty & \infty & \infty & 1 & 1 & 0
\end{bmatrix},
\qquad
C^{(3)} = \begin{bmatrix}
0 & 1 & 3 & 6 & 8 & \infty \\
1 & 0 & 2 & 5 & 7 & \infty \\
3 & 2 & 0 & 3 & 5 & \infty \\
6 & 5 & 3 & 0 & 1 & 1 \\
8 & 7 & 5 & 1 & 0 & 1 \\
\infty & \infty & \infty & 1 & 1 & 0
\end{bmatrix},
$$

$$
C^{(4)} = C^{(5)} = C^{(6)} = \begin{bmatrix}
0 & 1 & 3 & 6 & 7 & 7 \\
1 & 0 & 2 & 5 & 6 & 6 \\
3 & 2 & 0 & 3 & 4 & 4 \\
6 & 5 & 3 & 0 & 1 & 1 \\
7 & 6 & 4 & 1 & 0 & 1 \\
7 & 6 & 4 & 1 & 1 & 0
\end{bmatrix}.
$$

18.5 This matrix algorithm computes at the same time the costs of cheapest routes between all pairs of nodes. The algorithm is described in the following reference: Dantzig, G. B., All Shortest Routes in a Graph, *Technical Report No. 66-3*, Opns. Res. House, Stanford University (1966).

The successive matrices are

$$
C^{(1)} = [0], \qquad\qquad\qquad C^{(2)} = \begin{bmatrix} 0 & 1 \\ 1 & 0 \end{bmatrix},
$$

$$
C^{(3)} = \begin{bmatrix}
0 & 1 & 3 \\
1 & 0 & 2 \\
3 & 2 & 0
\end{bmatrix},
\qquad
C^{(4)} = \begin{bmatrix}
0 & 1 & 3 & 6 \\
1 & 0 & 2 & 5 \\
3 & 2 & 0 & 3 \\
6 & 5 & 3 & 0
\end{bmatrix},
$$

$$
C^{(5)} = \begin{bmatrix}
0 & 1 & 3 & 6 & 7 \\
1 & 0 & 2 & 5 & 6 \\
3 & 2 & 0 & 3 & 4 \\
6 & 5 & 3 & 0 & 1 \\
7 & 6 & 4 & 1 & 0
\end{bmatrix},
\qquad
C^{(6)} = \begin{bmatrix}
0 & 1 & 3 & 6 & 7 & 7 \\
1 & 0 & 2 & 5 & 6 & 6 \\
3 & 2 & 0 & 3 & 4 & 4 \\
6 & 5 & 3 & 0 & 1 & 1 \\
7 & 6 & 4 & 1 & 0 & 1 \\
7 & 6 & 4 & 1 & 1 & 0
\end{bmatrix}.
$$

18.6 The statement is false. It is possible for the cheapest path to go from n_1 to n_i and then *via* a path from n_i to n_j whose cost is cheaper than the cost of the link (n_i, n_j).

18.7 The trace of the cheapest route tree is

Link i	Predecessor P^*	Cost C^*
1	4	245
2	1	265
3	5	155
4	3	180
5	6	90
6	0	0
7	1	280

18.8 The algorithm gives $1, 4$ as the cheapest route with cost 6, instead of the route $1, 2, 3, 4$ with cost 5 units. The algorithm permanently labels node 3 first, which prevents consideration of the alternative route to 3 *via* 2. Although $1, 2, 3, 4$ is the cheapest route from 1 to 4, it is not true that $1, 2, 3$ is the cheapest route from 1 to 3.

18.9 (i) For each link $0 \leqslant f_i \leqslant u_i$ and at each node, Kirchhoff's law holds, with a flow value $g = 6$, and a network cost

$$C = \sum c_i f_i = 5 + 4 + 4 + 18 + 9 + 6 = 46.$$

(ii) With the same notation as in the Table 15.2 (Chap. III), we can tabulate the chains, chain flow, route costs, and the sums over dual variables as follows:

Chain	$\sum \mu_i$	$v - \sum \mu_i$	C_J	h_J
m_1	1	7	8	0
m_2	1	7	7	2
m_3	1	7	9	0
m_4	0	8	9	0
m_5	0	8	8	3
m_6	0	8	10	0
m_7	0	8	9	0
m_8	0	8	8	1
m_9	0	8	10	0

Also $V = vg - \sum \mu_i u_i = (8)(6) - 2 = 46.$

(iii) If $v^* - \sum \mu_i^* < C_j(O, D)$, then $h_j^* = 0$, verified for $j = 1, 3, 4, 6, 7, 9$;

If $\mu_i^* > 0$, then $f_i^* = u_i$, verified for $i = 3$;

If $h_j^* > 0$, then $v^* - \sum \mu_i^* = C_j(O, D)$; verified for $j = 2, 5, 8$;

If $f_i^* < u_i$, then $\mu_i^* = 0$, verified for all $i \neq 3$.

(iv) All chains are available for flow from chain m_2 and all have chain costs $\geqslant C_2 = 7$. All chains except m_1, m_2, m_3 are available for flow from chains m_5 and m_8 and have chain costs $\geqslant C_5 = C_8 = 8$.

18.10   (i) $C = \sum c_i f_i = 91$.

(ii)

Chain	$\sum \mu_i$	$v - \sum \mu_i$	C_J	h_J
m_1	2	8	8	0
m_2	3	7	7	2
m_3	1	9	9	0
m_4	1	9	9	0
m_5	2	8	8	4
m_6	1	9	10	0
m_7	1	9	9	1
m_8	2	8	8	2
m_9	0	10	10	2

Also $V = vg - \sum \mu_i u_i = (10)(11) - 6 - 5 - 8 = 91$.

(iii) If $v^* - \sum \mu_i^* < C_j(O, D)$, then $h_j^* = 0$, vertified for $j = 6$;

If $\mu_i^* > 0$, then $f_i^* = u_i$, verified for $i = 1, 6, 9$;

If $h_j^* > 0$, then $v^* - \sum \mu_i^* = C_j(O, D)$, verified for $j = 2, 5, 7, 8, 9$;

If $f_i^* < u_i$, then $\mu_i^* = 0$, verified for $i = 2, 5, 7, 8, 10$.

(iv) All chains except m_4 and m_5 are available for flow from chain m_1 and have chain costs $\geqslant C_2 = 7$. Chains m_4, m_6, and m_9 are available for flow from m_5, chain costs $\geqslant C_5 = 8$. Chain m_9 is available for flow from m_7, chain cost $\geqslant C_7 = 9$. Chains m_7 and m_9 are available for flow from m_8, chain costs $\geqslant C_8 = 8$. No chain is available for flow from m_9.

18.11 (i) The chains with route costs are

O–D pair	k	j	Chain $m_j^{(k)}$	Route cost $C_j^{(k)}$	Chain flow $h_j^{(k)}$
O_1–D_1	1	1	1, 6, 8, 10	7	0
		2	2, 4, 6, 8, 10	6	35
		3	2, 5, 7, 9, 10	8	0
O_1–D_2	2	1	1, 6, 8, 11	9	0
		2	2, 4, 6, 8, 11	8	20
		3	2, 5, 7, 9, 11	10	0
O_2–D_1	3	1	3, 4, 6, 8, 10	10	15
		2	3, 5, 7, 9, 10	12	0
O_2–D_2	4	1	3, 4, 6, 8, 11	12	30
		2	3, 5, 7, 9, 11	14	0

(ii) The all-or-nothing cheapest route assignment gives the chain flows $h_j^{(k)}$ listed above with consequent link flows:

Link	i	1	2	3	4	5	6	7	8	9	10	11
Commodity flow	$f_i^{(1)}$	0	35	0	35	0	35	0	35	0	35	0
	$f_i^{(2)}$	0	20	0	20	0	20	0	20	0	0	20
	$f_i^{(3)}$	0	0	15	15	0	15	0	15	0	15	0
	$f_i^{(4)}$	0	0	30	30	0	30	0	30	0	0	30
Link flow	f_i	0	55	45	100	0	100	0	100	0	50	50

Network cost is $C = 880$ units.

18.12

	Origin	Copy	Links in tree	Copy flow
	O_1	1	2, 4, 6, 8, 10, 11	$v^1 = 55$
	O_2	2	3, 4, 6, 8, 10, 11	$v^2 = 45$

Link	i	1	2	3	4	5	6	7	8	9	10	11
Copy flow	f_i^1	0	55	0	55	0	55	0	55	0	35	20
	f_i^2	0	0	45	45	0	45	0	45	0	15	30
Link flow	f_i	0	55	45	100	0	100	0	100	0	50	50

18.13

Destination	Copy	Links in tree	Copy flow
D_1	1	2, 3, 4, 6, 8, 10	$v^1 = 50$
D_2	2	2, 3, 4, 6, 8, 11	$v^2 = 50$

Link	i	1	2	3	4	5	6	7	8	9	10	11
Copy flow	f_i^1	0	35	15	50	0	50	0	50	0	50	0
	f_i^2	0	20	30	50	0	50	0	50	0	0	50
Link flow	f_i	0	55	45	100	0	100	0	100	0	50	50

18.14   With the chains enumerated as in the answer to Problem 18.11, the chain flows for the two loadings (i), (ii) are

O–D pair	k	j	Chain flows (i)	$h_j^{(k)}$ (ii)
O_1–D_1	1	1	0	35
		2	35	0
		3	0	0
O_1–D_2	2	1	5	15
		2	15	5
		3	0	0
O_2–D_1	3	1	0	15
		2	15	0
O_2–D_2	4	1	0	30
		2	30	0

The resulting link flows are

Link	i	1	2	3	4	5	6	7	8	9	10	11
(i) Link flow	f_i	5	50	45	50	45	55	45	55	45	50	50
(ii) Link flow	f_i	50	5	45	50	0	100	0	100	0	50	50

Network costs are (i) $C = 975$, (ii) $C = 930$. Both these network flows satisfy the first extremal principle.

18.15   In the notation of (72)–(75), Sect. 16, Chap. III, and the answer to Problem 18.11, the problem of minimizing the network cost can be expressed as the LP

$$h_1^{(1)}, h_2^{(1)}, h_3^{(1)}, h_1^{(2)}, h_2^{(2)}, h_3^{(2)}, h_1^{(3)}, h_2^{(3)}, h_1^{(4)}, h_2^{(4)} \geqslant 0,$$

$$h_1^{(1)} + h_2^{(1)} + h_3^{(1)} = 35,$$

$$h_1^{(2)} + h_2^{(2)} + h_3^{(2)} = 20,$$

$$h_1^{(3)} + h_2^{(3)} = 15,$$

$$h_1^{(4)} + h_2^{(4)} = 30,$$

$$h_2^{(1)} + h_2^{(2)} + h_1^{(3)} + h_1^{(4)} + s = 50,$$

$$7h_1^{(1)} + 6h_2^{(1)} + 8h_3^{(1)} + 9h_1^{(2)} + 8h_2^{(2)} + 10h_3^{(2)}$$

$$+ 10h_1^{(3)} + 12h_2^{(3)} + 12h_1^{(4)} + 14h_2^{(4)} = C(\min).$$

The quantity s is a nonnegative slack flow. By means of any of the standard LP algorithms, this problem can be expressed as

$$h_1^{(1)} = 35 - h_2^{(1)} - h_3^{(1)},$$

$$h_1^{(2)} = 15 + h_2^{(1)} - h_3^{(2)} - h_2^{(3)} - h_2^{(4)} + s,$$

$$h_2^{(2)} = 5 + h_2^{(1)} + h_3^{(2)} + h_2^{(4)} - s,$$

$$h_1^{(3)} = 15 - h_2^{(3)},$$

$$h_1^{(4)} = 30 - h_2^{(4)},$$

$$C = 930 + h_3^{(1)} + h_3^{(2)} + h_2^{(3)} + h_2^{(4)} + s.$$

The optimal solution which minimizes C is read off as $h_1^{(1)} = 35$, $h_1^{(2)} = 15$, $h_2^{(2)} = 5$, $h_1^{(3)} = 15$, $h_1^{(4)} = 30$ with all other chain flows zero, and $C^* = 930$. This is the solution for Case (ii) in the previous problem.

18.16 With the same notation as in the answer to Problem 18.11, the nonzero chain flows for the user-optimized traffic pattern are:

	$m_j^{(k)}$	$C_j^{(k)}$	$h_j^{(k)}$
O_1–D_1	$m_1^{(1)}$	14	h
O_1–D_1	$m_3^{(1)}$	14	$35-h$
O_1–D_2	$m_1^{(2)}$	16	$40-h$
O_1–D_2	$m_3^{(2)}$	16	$h-20$
O_2–D_1	$m_2^{(3)}$	16	15
O_2–D_2	$m_2^{(4)}$	18	30

The flow on chain $m_1^{(1)}$ can be taken equal to any value h, $0 \leqslant h \leqslant 20$. The optimal chain flow pattern is therefore not unique but the unique optimal link flow pattern is:

$$f_1{}^* = 40, \quad f_2{}^* = 15, \quad f_3{}^* = 45, \quad f_4{}^* = 0, \quad f_5{}^* = 60,$$

$$f_6{}^* = 40, \quad f_7{}^* = 60, \quad f_8{}^* = 40, \quad f_9{}^* = 60, \quad f_{10}^* = 50, \quad f_{11}^* = 50.$$

18.17   Using the same notation as in the answer to Problem 18.11 the nonzero chain flows for the system-optimized traffic pattern are:

	$m_j^{(k)}$	$C_j^{(k)}$	$h_j^{(k)}$
O_1–D_1	$m_1^{(1)}$	10	h
O_1–D_1	$m_3^{(1)}$	15	$35-h$
O_1–D_2	$m_1^{(2)}$	12	$30-h$
O_1–D_2	$m_3^{(2)}$	17	$h-10$
O_2–D_1	$m_2^{(3)}$	17	15
O_2–D_2	$m_2^{(4)}$	19	30

The flow on chain $m_1^{(1)}$ can be taken equal to any value h, $0 \leqslant h \leqslant 10$. The optimal chain flow pattern is therefore not unique but the unique optimal link flow pattern is:

$$f_1^* = 30, \quad f_2^* = 25, \quad f_3^* = 45, \quad f_4^* = 0, \quad f_5^* = 70,$$

$$f_6^* = 30, \quad f_7^* = 70, \quad f_8^* = 30, \quad f_9^* = 70, \quad f_{10}^* = 50, \quad f_{11}^* = 50.$$

18.18   Using Appendix B we can write the maximal flow problem as the following primal linear program, with its dual obtained by using dual variables $-\lambda_i$ for $i = 1, 2, \ldots, n$ and μ_{ij} for $(i,j) \in L$:

Primal P	Dual D
$f_{n1} \geqslant 0$	$\lambda_n - \lambda_1 \geqslant 1$
$f_{ij} \geqslant 0$	$\mu_{ij} + \lambda_i - \lambda_j \geqslant 0 \qquad (i,j) \in L$
$f_{ij} \leqslant u_{ij}$	$\mu_{ij} \geqslant 0 \qquad (i,j) \in L$
$\sum_{A(1)} f_{1j} - f_{n1} = 0$	λ_1 unrestricted in sign
$\sum_{A(i)} f_{ij} - \sum_{B(i)} f_{ji} = 0$	λ_i unrestricted in sign $i = 2, \ldots, n-1$
$f_{n1} - \sum_{B(n)} f_{jn} = 0$	λ_n unrestricted in sign
$-f_{n1} = z(\min)$	$-\sum_{(i,j)\in L} u_{ij}\mu_{ij} = v(\max).$

For any cut-set $(X, \overline{X})$ separating the origin from the destination,

$$\lambda_i = \begin{cases} 0 & \text{if } i \in X \\ 1 & \text{if } i \in \overline{X} \end{cases}$$

and

$$\mu_{ij} = \begin{cases} 1 & \text{if}(i,j) \in (X, \overline{X}) \\ 0 & \text{otherwise} \end{cases}$$

give a feasible solution of the dual with

$$v = -u(X, \overline{X}).$$

By (6) in Appendix B,

$$z = -f_{n1} \geqslant v = -u(X, \overline{X})$$

or, since $f_{n1} = g = $ flow value,

$$g \leqslant u(X, \overline{X})$$

as given in (10), Sect. 9, Chap. II. For optimal primal and dual solutions,

$$f_{n1}^* = \sum u_{ij} \mu_{ij}^*,$$

or

$$g^* = u(X^*, \overline{X}^*)$$

which proves the max-flow min-cut theorem.

The complementary slackness relations are:

if $\lambda_n^* - \lambda_1^* > 1$ then $f_{n1}^* = 0$;

if $\mu_{ij}^* + \lambda_i^* - \lambda_j^* > 0$ then $f_{ij}^* = 0$ $(i,j) \in L$;

if $\mu_{ij}^* > 0$ then $f_{ij}^* = u_{ij}$ $(i,j) \in L$;

if $f_{n1}^* > 0$ then $\lambda_n^* - \lambda_1^* = 1$,

if $f_{ij}^* > 0$ then $\mu_{ij}^* + \lambda_i^* - \lambda_j^* = 0$,

if $f_{ij}^* < u_{ij}$ then $\mu_{ij}^* = 0$.

The dependence of the optimal dual variables on the optimal link flows can be expressed as follows:

$$\lambda_1^* - \lambda_n^* \leqslant -1 \quad \text{for} \quad f_{n1}^* = 0,$$

$$\text{and} \quad \lambda_1^* - \lambda_n^* = -1 \quad \text{for} \quad f_{n1}^* = 0;$$

$$\lambda_j^* - \lambda_i^* \leqslant 0 \quad \text{for} \quad f_{ij}^* = 0,$$

$$\lambda_j^* - \lambda_i^* = 0 \quad \text{for} \quad 0 < f_{ij}^* < u_{ij},$$

$$\lambda_j^* - \lambda_i^* \geqslant 0 \quad \text{for} \quad f_{ij}^* = u_{ij};$$

$$\mu_{ij}^* = 0 \quad \text{for} \quad f_{ij}^* < u_{ij},$$

$$\mu_{ij}^* \geqslant 0 \quad \text{for} \quad f_{ij}^* = u_{ij}.$$

Compare Fig. 15.1, Chap. III.

Chapter IV

27.1 From (13), Sect. 21,

$$c'_{13} = \frac{2(5-5)+6(5-3)}{2+6} = \frac{12}{8},$$

$$c'_{23} = \frac{2(9-5)+6(9-3)}{2+6} = \frac{44}{8},$$

$$c'_{33} = \frac{2(7-5)+6(7-3)}{2+6} = \frac{28}{8}.$$

The new table of costs with the new values a_i' and b_j' is

			a_i'
5	4	1.5	5
10	8	5.5	5
9	9	3.5	5

$$b_j' \quad 1 \quad 6 \quad 8$$

By (18), Sect. 21,

$$\beta_3' = \frac{2(-5)+6(-3)}{2+6} = -\frac{28}{8},$$

giving a compressed trip table with new implicit prices

			α_i'
1	3	1	5
0	3	2	9
0	0	5	7

$$\beta_j' \quad 0 \quad -1 \quad -3.5$$

The optimality of these flows is verified from the complementary slackness relations:

$$f'_{11} = 1, \qquad \alpha_1' + \beta_1' = 5 = c'_{11},$$

$$f'_{12} = 3, \qquad \alpha_1' + \beta_2' = 4 = c'_{12},$$

$$f'_{13} = 1, \qquad \alpha_1' + \beta_3' = 1.5 = c'_{13},$$

$$f'_{22} = 3, \qquad \alpha_2' + \beta_2' = 8 = c'_{22},$$

$$f'_{23} = 2, \qquad \alpha_2' + \beta_3' = 5.5 = c'_{23},$$

$$f'_{33} = 5, \qquad \alpha_2' + \beta_3' = 3.5 = c'_{33},$$

$$\alpha_2' + \beta_1' = 9 < c'_{21} = 10, \qquad f'_{21} = 0,$$

$$\alpha_3' + \beta_1' = 7 < c'_{31} = 9, \qquad f'_{31} = 0,$$

$$\alpha_3' + \beta_2' = 6 < c'_{32} = 9, \qquad f'_{32} = 0.$$

It is interesting to note that the trip table

$$\begin{array}{ccc|}
1 & 4 & 0 \\
0 & 2 & 3 \\
0 & 0 & 5
\end{array}$$

with $mn - (m+n-1) = 4$ zero entries, is another optimal solution for the new Hitchcock problem, but it is not obtained by compressing the original table.

27.2   From (12), Sect. 21,

$$c'_{11} = \frac{1 \times 5 + 6 \times 4}{1+6} = \frac{29}{7}, \qquad c'_{31} = \frac{1 \times 9 + 6 \times 9}{1+6} = 9,$$

and from (13), Sect. 21,

$$c'_{21} = \frac{1(9+0) + 6(9-1)}{1+6} = \frac{57}{7},$$

so that the new cost table is

$$a_i'$$

$\dfrac{29}{7}$	3	2	5
$\dfrac{57}{7}$	4	7	5
9	8	4	5

$$b_j' \qquad 7 \quad 2 \quad 6$$

By (18), Sect. 21,

$$\beta_1' = \frac{1(0)+6(-1)}{1+6} = -\frac{6}{7}.$$

The compressed trip table with new implicit prices is

$$
\begin{array}{ccc|c}
4 & 0 & 1 & 5 \\
3 & 2 & 0 & 9 \\
0 & 0 & 5 & 7 \\
\hline
-\dfrac{6}{7} & -5 & -3 &
\end{array}
$$

The complementary slackness relations are:

$$f_{11}' = 4, \qquad \alpha_1' + \beta_1' = \frac{29}{7} = c_{11}',$$

$$f_{13}' = 1, \qquad \alpha_1' + \beta_3' = 2 = c_{13}',$$

$$f_{21}' = 3, \qquad \alpha_2' + \beta_1' = \frac{57}{7} = c_{21}',$$

$$f_{22}' = 2, \qquad \alpha_2' + \beta_2' = 4 = c_{22}',$$

$$f_{33}' = 5, \qquad \alpha_3' + \beta_3' = 4 = c_{33}',$$

$$\alpha_2' + \beta_3' = 6 < c_{23}' = 7, \qquad f_{23}' = 0,$$

$$\alpha_3' + \beta_1' = \frac{43}{7} < c_{31}' = 9, \qquad f_{31}' = 0,$$

$$\alpha_3' + \beta_2' = 2 < c_{32}' = 8, \qquad f_{32}' = 0.$$

27.3   From (27), Sect. 22, the trip table is calculated to be

$$
\begin{array}{c}
 \\
\begin{bmatrix}
4 & 8 & 12 & 16 \\
3 & 6 & 9 & 12 \\
2 & 4 & 6 & 8 \\
1 & 2 & 3 & 4
\end{bmatrix}
\end{array}
\begin{array}{c}
a_i \\
40 \\
30 \\
20 \\
10
\end{array}
$$

$$b_j \quad 10 \quad 20 \quad 30 \quad 40 \quad 100 = v$$

Combining centroids 2 and 3 gives

$$
\begin{array}{c}
 & & & a_i' \\
\begin{bmatrix} 4 & 20 & 16 \\ 5 & 25 & 20 \\ 1 & 5 & 4 \end{bmatrix} & \begin{matrix} 40 \\ 50 \\ 10 \end{matrix}
\end{array}
$$

$$b_j' \quad 10 \quad 50 \quad 40 \quad 100 = v'$$

while removing centroid 2 gives

$$
\begin{array}{c}
 & & & a_i' \\
\begin{bmatrix} 4 & 12 & 16 \\ 2 & 6 & 8 \\ 1 & 3 & 4 \end{bmatrix} & \begin{matrix} 32 \\ 16 \\ 8 \end{matrix}
\end{array}
$$

$$b_j' \quad 7 \quad 21 \quad 28 \quad 56 = v'$$

27.4   The following results were obtained by computer calculations:

$$
f_{ij} = \begin{bmatrix} 0.37 & 2.20 & 0.55 & 1.88 \\ 0.30 & 2.17 & 1.04 & 1.49 \\ 0.33 & 1.63 & 0.41 & 2.63 \end{bmatrix}.
$$

The values

$$x_1 = 0.616, \qquad x_2 = 1.419, \qquad x_3 = 1.312,$$

$$y_1 = 0.115, \qquad y_2 = 0.554, \qquad y_3 = 0.113, \qquad y_4 = 0.310,$$

$$\gamma = 0.211,$$

were obtained at the end of the iterations.

27.5   The design year productions and attractions are

	Design year	
Zone i	Production a_i	Attraction b_i
1	14	33
2	33	28
3	28	14

Consider the iterative scheme

$$f_{ij}^{(k+1)} = x_i^{(k)} f_{ij}^{(k)} y_j^{(k)},$$

$$y_j^{(k)} = b_j \left[\sum_i f_{ij}^{(k)} \right]^{-1},$$

$$x_i^{(k)} = a_i \left[\sum_j f_{ij}^{(k)} y_j^{(k)} \right]^{-1},$$

for $k = 1, 2, \ldots$. Applying these to the given data yields the successive iterates:

$$[f_{ij}^{(2)}] = \begin{bmatrix} 2 & 10 & 2 \\ 19 & 8 & 6 \\ 16 & 6 & 6 \end{bmatrix} \begin{matrix} 14 \\ 33 \\ 28 \end{matrix} ,$$
$$\qquad\qquad 37 \quad 24 \quad 14$$

$$[f_{ij}^{(3)}] = \begin{bmatrix} 2 & 10 & 2 \\ 17 & 10 & 6 \\ 15 & 7 & 6 \end{bmatrix} \begin{matrix} 14 \\ 33 \\ 28 \end{matrix} ,$$
$$\qquad\qquad 34 \quad 27 \quad 14$$

$$[f_{ij}^{(4)}] = \begin{bmatrix} 2 & 10 & 2 \\ 16 & 11 & 6 \\ 15 & 7 & 6 \end{bmatrix} \begin{matrix} 14 \\ 33 \\ 28 \end{matrix} .$$
$$\qquad\qquad 33 \quad 28 \quad 14$$

Calculations have been approximate and answers rounded off to give integers. This simple growth factor and scaling technique was developed by Fratar as one of the earliest trip distribution models. This and the next two problems, together with the answers, have been taken from the Manual for the Urban Planning System/360 Trip Distribution Programs, Bureau of Public Roads, US Department of Transportation, 1968.

27.6   The successive values of b_j are

	1	2	3
		j	
$b_j^{(1)}$	33	28	14
$b_j^{(2)}$	28	33	14
$b_j^{(3)}$	27	34	13

with corresponding trip tables

$$[f_{ij}^{(1)}] = \begin{bmatrix} 2 & 10 & 2 \\ 19 & 8 & 6 \\ 17 & 5 & 6 \end{bmatrix} \begin{matrix} 14 \\ 33 \\ 28 \end{matrix}$$
$$\qquad\quad 38 \quad 23 \quad 14$$

$$[f_{ij}^{(2)}] = \begin{bmatrix} 2 & 10 & 2 \\ 17 & 10 & 6 \\ 15 & 7 & 6 \end{bmatrix} \begin{matrix} 14 \\ 33 \\ 28 \end{matrix}$$
$$\qquad\quad 34 \quad 27 \quad 14$$

$$[f_{ij}^{(3)}] = \begin{bmatrix} 1 & 11 & 2 \\ 17 & 10 & 6 \\ 15 & 7 & 6 \end{bmatrix} \begin{matrix} 14 \\ 33. \\ 28 \end{matrix}$$
$$\qquad\quad 33 \quad 28 \quad 14$$

Calculations have been approximate and answers rounded off to give integers.

27.7   The successive values of b_j are

	1	2	3
		j	
$b_j^{(1)}$	33	28	14
$b_j^{(2)}$	35	41	13
$b_j^{(3)}$	37	48	14

with corresponding trip tables

$$[f_{ij}^{(1)}] = \begin{bmatrix} 10 & 2 & 1 \\ 9 & 14 & 2 \\ 12 & 3 & 12 \end{bmatrix} \begin{matrix} 13 \\ 25, \\ 27 \end{matrix}$$

$$\qquad\qquad 31 \quad 19 \quad 15 \qquad 65$$

$$[f_{ij}^{(2)}] = \begin{bmatrix} 11 & 3 & 0 \\ 7 & 18 & 2 \\ 13 & 3 & 11 \end{bmatrix} \begin{matrix} 14 \\ 27, \\ 27 \end{matrix}$$

$$\qquad\qquad 31 \quad 24 \quad 13 \qquad 68$$

$$[f_{ij}^{(3)}] = \begin{bmatrix} 11 & 1 & 1 \\ 7 & 20 & 2 \\ 12 & 3 & 12 \end{bmatrix} \begin{matrix} 13 \\ 29. \\ 27 \end{matrix}$$

$$\qquad\qquad 30 \quad 24 \quad 15 \qquad 69$$

Calculations have been approximate and answers rounded off to give integers.

27.8   The iterative steps and the destination-optimal distribution are obtained as follows:

1	2	3	4	5	6	7
7	2	1	7	5	2	4
6	2	1	7	5	2	4
6	2	1	7	5	1	4
6	2	1	7	5	3	4

27.9   (a)   We first find stationary points of

$$\mathcal{L}(\rho_{ij}, \lambda_i, \mu_j, \delta, \nu) = -\sum_{i,j} \rho_{ij} \ln \rho_{ij} - \sum_i \lambda_i \left(\sum_j \rho_{ij} - u_i \right)$$

$$- \sum_j \mu_j \left(\sum_i \rho_{ij} - v_j \right) - \delta(\rho_{11} - k) - \nu \left(\sum_{i,j} \rho_{ij} - 1 \right).$$

The partial derivatives of $\mathscr{L}$ with respect to ρ_{ij} are

$$\frac{\partial \mathscr{L}}{\partial \rho_{ij}} = -1 - \ln \rho_{ij} - \lambda_i - \mu_j - \nu \qquad i,j \neq 1,1,$$

$$= -1 - \ln \rho_{ij} - \lambda_i - \mu_j - \delta - \nu \qquad i,j = 1,1.$$

The optimal flows can, therefore, be rewritten in the form

$$\rho_{ij}^* = x_i y_j \qquad i,j \neq 1,1,$$

$$= \phi x_1 y_1 \qquad i,j = 1,1,$$

where $\phi = e^{-\delta}$ and row and column sums are

$$\sum_{j=1}^{n} \rho_{ij}^* = x_i \left(\sum_{j=1}^{n} y_j \right) \qquad\qquad = u_i, \qquad i \neq 1$$

$$= x_1 (\phi y_1 + y_2 + \ldots + y_n) = u_1, \qquad i = 1,$$

$$\sum_{i=1}^{m} \rho_{ij}^* = \left(\sum_{i=1}^{m} x_i \right) y_j \qquad\qquad = v_j, \qquad j \neq 1$$

$$= (\phi x_1 + x_2 + \ldots + x_m) y_1 = v_1, \qquad j = 1.$$

The ratio of row totals is

$$u_i / u_2 = x_i / x_2, \qquad i \neq 1.$$

Similarly,

$$v_j / v_2 = y_j / y_2, \qquad j \neq 1.$$

The optimal solution for ρ_{ij}^* can be written in terms of these ratios and the given value of $\rho_{11} = k$. We obtain

$$k + \rho_{12}^* \frac{1 - v_1}{v_2} = u_1,$$

$$k + \rho_{21}^* \frac{1 - u_1}{u_2} = v_1,$$

$$\rho_{12}^* + \rho_{22}^* \frac{1 - v_1}{v_2} = u_2.$$

Thus,

$$\rho_{12}^* = \frac{v_2(u_1 - k)}{1 - v_1},$$

$$\rho_{21}^* = \frac{u_2(v_1 - k)}{1 - u_1},$$

$$\rho_{22}^* = \frac{u_2 v_2 (1 - u_1 - v_1 + k)}{(1 - u_1)(1 - v_1)},$$

$$\rho_{ij}^* = \frac{v_j}{v_2}\rho_{i2}^* = \frac{u_i}{u_2}\rho_{2j}^*, \qquad i,j \neq 1, 2.$$

The computation proceeds by finding ρ_{12}^*, ρ_{21}^*, ρ_{22}^*, ρ_{1j}^*, ρ_{2j}^* and then all other ρ_{ij}^*.

(b) For $\rho_{11} = 0.01$, the optimal solution is

			u_i
0.01	0.08	0.16	0.25
0.13	0.04	0.08	0.25
0.26	0.08	0.16	0.50

| v_j | 0.40 | 0.20 | 0.40 |

AUTHOR INDEX

Numbers in parentheses are reference numbers and indicate that an author's work is referred to, although his name is not cited in the text. Numbers in italics show the page on which the complete reference is listed.

B

Bay Area Transportation Study Commission, 5(1), 13(1), *14*

Beckmann, M. J., 17(6), *45, 108,* 145

Bellman, R., 51, *103*

Berge, C., 17(2), *44*

Blackburn, J. B., *109*

Blunden, W. R., 144(15), *148*

Brokke, G. E., 64(8), *103*

Burrell, J. E., 65, *105,* 142(17), *149*

Busacker, R. G. 17(3), *44*

C

Caldwell, T., 57, 61, 62, *103*

Charnes, A., 100, *108,* 149

Control Data Corporation, *104*

Cooper, W. W., 100, *108,* 149

D

Dafermos, S. C., *107*

Dantzig, G. B., 66(18), 67(18), *106,* 107, 119(18), *170*

Dearinger, J. A., 144(13), *148*

Dijkstra, E. W., 52(6), *103*

Dreyfus, S. E., 51, *102*

F

Fairthorne, D. B., 144(14), *148*

Floyd, R. W., *169*

Ford, L. R., 17(1), 43(1), *44,* 75(1), 107, 119(1)

Fulkerson, D. R., 17(1), 43(1), *44,* 75(1), *107,* 119(1)

Funk, M. L., *109*

G

Gale, D., 138 (11), *147*

Ghouila-Houri, A., 17(2), *44*

Gibert, A., 101(27), *109*

Golob, T. F., 145

Greater London Council, 10(2), 13(2), *14*

H

Harary, F., 17(5), *44*

Heanue, K. E., 134(12), 144(12), *147*

Heggie, I. G., 136(8), 144(8), *146,* 147

187

Highway Research Board, 41, *45*
Hitchcock, F. L., 116, 118– 121, *144*
Hoffman, W., *168*

I

Irwin, N. A., 142(16), *148*

J

Jansen, G. R., 63, *105*
Jorgenson, N. O., 66(17), 67(17), 70, 96(17), *106*

K

Kaufmann, A., 17(4), *44*
Kirby, R. F., 51(4), 57(4), 63(11), *103*, *104*, 138, 141(10), *147*
Kitchen, J. W., *108*, 160(23)

L

Lawson, M. C., 144(13), *148*

M

McDonald, W. R., 144(15), *148*
McGuire, C. B., 17(6), *45*
Metropolitan Corporation of Greater Winnipeg, 13(3), *14*
Meyers, D. A., 142(17), *149*
Michaels, R. M., 65, *105*
Murchland, J. D., 51, *102*, 122, *144*

N

Nash, J., *107*

P

Pavley, R., *168*
Pinnell, C., 101, *109*
Potts, R. B., 51(4), 57(4), *103*

Powell, T J., 142(17), *149*
Pyers, C. E., 134(12), 144(12), *147*

S

Saaty, T. L., 17(3), *44*
Sasaki, T., 131, 133(4), *145*
Satterly, G. T., 101, *109*
Sema, Group Metra France, 62(10), *104*
Shapley, L. S., 138(11), *147*
Sinkhorn, R., 133, 135(6), *145*
Smith, W. S., 13(4), *14*
Snell, R. R., *109*
Sparrow, F. T., *107*
Stouffer, S. A., 136, *147*
Stover, V. G., 64(14), *105*

T

Tanner, J. C., *146*
Tomlin, J. A., 101, *109*, 131, 133(5), 143(18), *145*, *149*
Tomlin, S. G., 131, 133(5), *145*
Traffic Research Corporation, 142
Tresidder, J. O., 142(17), *149*

U

U.S. Department of Transportation, *182*

V

Veinott, A. F., 106
Von Cube, A. G., 142(16), *148*

W

Wachs, M., 63(12), *104*
Wardrop, J. G., 49, *102*
White, D. J., 145
Wilson, A. G., 122(3), *144*, 146, 147
Winsten, C. B., 17(6), *45*

SUBJECT INDEX

Numbers in italics indicate the first occurrence or formal definition of a term.

A

Accessible node, 22
Admissible route, *61*
After nodes, *27*
Algorithms
  gravity model, 135–136
  intervening opportunities, 138
  matrix scaling, 145–146
  mean trip length model, 132–133
  out-of-kilter, *66*, 75–86
  preferencing model, 140–141
  shortest path, 102–104
  tree-building, 52–56, 60
All-or-nothing assignment, *64*
A-node, *18*
Arborescence, *26*
Assignment
  all-or-nothing, *64*
  associated traffic, *95–96*
  cheapest route, 63–64
  combined distribution, 116, 141–143
  congested, *100*–101
  diversion, *64*
  future traffic, *13*

multicommodity distribution–assignment, 143
  multiple route, *65*
  traffic, 12, 115–116
Attraction nodes, 27–29
Available chains, *70*
Average costs of chain flow, 95, 159–162

B

Base year inventory, 11–12
Bipartite graph, *18*
B-node, *18*

C

Capacitated network, *41*–43
Centroid, *4*, 26–29, 115–116
Chain, *20*, 21–22
  available, *70*
  flow, 71
  traffic pattern, *33*
  unavailable, *70*
Cheapest route assignment, *63–64*
City street network, *2*, 3, 40

189

Coding a network, *5*
Commodity, *34*, 96
Complementary slackness, 67, 76, *158*, 177–178
Complete graph, *18*
    nonplanar, *9*
Compressed network, *36*–37
Compressibility, *36*, 120–121, 128–130, 136
Congested assignment, *100*–101
Connected directed graph, *22*
Connected nodes, *22*
Conservation equations, 27, 30–31, 108
Conservation principle, 26–38
Copy, *34*
Cordon line, *6*
Costs, *38*
    link, *38*
    flow independent, *39*
    route, *39*
Cut capacity, *42*, 43
Cut-set, *23*, 24
Cycle, 19, *21*

D
Dearest path, *110*
Destination node, *20*, 29
Destination-optimal, *139*
Deterrence function, *134*–135, 141–142
Directed graph, *18*
Directed link, *2*
Distribution
    combined with assignment, 116, 141–143
    compressibility, 117, 120–121, 128–130, 136
    destination-optimal, *139*
    entropy, 116, 123–128
    equilibrium, 122
    formulation, 116–118
    mean trip length, *130*
    gravity, 116, 133–136
    origin-optimal, *139*
    preferencing, 116, 138–141
    proportional, 123–128
    trip, 12, 115–116
Districts, *4*
Diversion assignment, *64*

Dual linear program, 67, 119, 121, *157–158*, 176–177

E
Economic analysis, *13*
Enroute points, *26*
Entropy models, 121–136
Expanded network, *36*, 38
Extremal principles, *50*, 68

F
Flow independent link cost, *39*
Flow value, *30*, 32
Flow-augmenting path, *78*
Flows
    chain, 32–36, 70–75
    commodity, *34*
    copy, *34*, 35
    link, 66–71
    network, 42
Forward link, *21*
Future traffic analysis, *13*

G
Graphs, 8
    accessible, *22*, 23
    bipartite graph, *18*
    complete, *18*
    connected directed, 22
    mixed, 24
    partial, *18*
    subgraph, *18*
    undirected, *24*
Gravity model, 122, 133–136

H
Hitchcock model, 118–121
Home node, *25*, 110

I
Interactance model, 134
Intermediate node, *4*, 27
Intervening opportunities model, 137–138
Inventory
    of main roads and transit services, 12
    of planning factors, 12
    of travel patterns, 12

J

Joined nodes, 18

K

Kilter numbers, 76, *77*
Kirchhoff's law, *26–27*, 29, 36, 38

L

Labels, 55
Lagrange function, 123, 134
Lagrange multiplier, 123–124, 134
Linear programming
    complementary slackness, 67, 76, *158*, 177–178
    dual, 67, 119, 121, *157–158*, 176–177
    primal, 67, *157–158*, 176–177
Link, 2, *18*
    capacity, *41*
    dummy, *4–5*
    kilter numbers, 77
    saturated, *70*
    unsaturated, *70*, 72
    in-kilter, 76–77
    out-of-kilter, 76–77
Link–chain incidence matrix, *33*
Link cost, *38*
Link flow traffic pattern, *32*
Loop, *18*
LTS program, 142–143

M

Main road network, 4, 5–7
Marginal costs of chain flow, 91, 95, 159–162
Matrix
    distribution, 115
    link–chain incidence, *33*, 35, 45
    node–link incidence, *31*, 34, 45
    scaling theorem, *145*
Mean trip length, *130*
    model, 130–133
Mesh, 21, *22*
Minimum network cost, 65–75
Mixed graph, *24*
Models, 115–144
    entropy, 121–136
    gravity, 122, 133–136
    Hitchcock, 118–121

    interactance, 134
    intervening opportunities, 137–138
    mean trip length, 130–133
    opportunity, 136–141
    preferencing, 138–141
    proportional, 123–130
Multicommodity distribution assignment, 143
Multiple O–D network, 34–36
Multiple route assignment, *65*

N

Nash equilibria, *107*
Network
    capacitated, *41*–43
    city street, *2*, 3, 40
    complete nonplanar, *9*
    compressed, *36*–37
    cost, *40*, 65–101
    entropy, 121–136, *122*
    evaluation, *13*
    expanded, *36*, 38
    feasible flow, *42*
    flow value, *30*, 32
    main road, *4*, 5–7
    minimum cost, 65–75
    multiple O–D, 34–36
    with prohibited turn, 60
    pseudo-, 57, 61, 62
    single O–D, *29*, 30–34
    spider web, 9
    traffic desire, 8–9
    transportation, 1, *2*
    with turn penalties, 56–62
Node–link incidence matrix, *31*
Nodes, *18*, 2
    accessible, 22
    attraction, 27–29
    centroid, 26–29
    connected, 22
    destination, *20*, 29
    distinct, *60*
    home, *25*, 110
    intermediate, *4*, 27
    joined, 18
    labelled, 55
    origin, *20*, 29
    predecessor, 154–155

production, 27–29
sink, 27
source, 27

O

Opportunity models, 136–141
Origin node, 20, 29
Origin-optimal, 139
Out-of-kilter, 75, 76–77

P

Partial graph, 18
Path, 21
Planning factors, 12
Preferencing model, 138–141
Primal linear program, 67, 157–158, 176–177
Principles
    available chains, 50, 75
    compressibility, 36
    conservation, 26–38
    equilibrium distribution, 122
    extremal, 50
    Kirchhoff's law, 26–27, 29, 36, 38
    Maximum entropy, 122, 123
    minimum network cost, 91–95
    separability, 36
    system-optimized traffic patterns, 50
    user-optimized traffic patterns, 50, 89
    Wardrop's, 49–50
Production nodes, 27–29
Proportional model, 123–130
Pseudo-network, 57, 61–62

R

Reverse link, 21
Route, 17
    admissible, 61
    cost, 39–40, 70

S

Saturated link, 70
Screen line, 6
Sectors, 4
Separability, 36, 38
SHARE program, 106
Single O–D network, 29, 30–34
Sink, 27
Skim tree, 55, 116

Source, 27
Spanning tree, 24
Spider web network, 9
Subgraph, 18
System-optimized traffic pattern, 50

T

Theorems
    cheapest route, 51, 153
    matrix scaling, 145–146
    max-flow, min-cut, 43
    net flow, 30
    system-optimized traffic patterns, 91–95
    turn penalty, 59
    user-optimized traffic patterns, 90
Total link cost, 38
Traffic desire 9, 115
    network, 8–9
Transportation
    networks, 1–10, 2, 13
    planning, 10–13, 115–117
    studies, 5, 6, 13–15, 31, 137, 142, 148
Travel forecasts, 12
Travel patterns, 12
TRC program, 142
Tree, 24
    -building algorithms, 52–56, 60, 169–171
    skim, 55, 116
    spanning, 24
    trace, 55–56, 171
Trip distribution, 9, 12, 115–152
Trip generation, 12
Trip length frequency distributions, 130
Turn penalties, 56

U

Unavailable chains, 70
Undirected graph, 24
Undirected link, 2, 24
Unsaturated link, 70, 72
User-optimized traffic pattern, 50, 89

Z

Zones, 4
    destination, 115
    origin, 115